OCR

Biology AS

Student Workbook

Model Answers: 2009

This model answer book is a companion publication to provide answers for the exercises in the **OCR Biology AS Student Workbook** 2009 edition. These answers have been produced as a separate publication to keep the cost of the workbook itself to a minimum, as well as to prevent easy access to the answers by students. In most cases, simply the answer is given with no working or calculations described. A few, however, have been provided with more detail because of their difficult nature.

www.biozone.co.uk

ISBN 978-1-877462-17-7

Published by BIOZONE International Ltd

BIOZONE

Additional copies of this Model Answers book may be purchased directly from the publisher

EUROPE & MIDDLE EAST:
BIOZONE Learning Media (UK) Ltd
P.O. Box 23698, Edinburgh
EH5 2WX, Scotland
Telephone: 131-557-5060
FAX: 131-557-5030
E-mail: sales@biozone.co.uk

ASIA & AUSTRALIA:
BIOZONE Learning Media Australia
P.O. Box 7523,
GCMC 4217 QLD, Australia
Telephone: +61 7-5575-4615
FAX: +61 7-5572-0161
E-mail: sales@biozone.com.au

NORTH & SOUTH AMERICA, AFRICA:
BIOZONE International Ltd
P.O. Box 13-034,
Hamilton, New Zealand
Telephone: +64 7-856-8104
FAX: +64 7-856-9243
FREE FAX: 1-800-717-8751(USA/Canada only)
E-mail: sales@biozone.co.nz

Contents

Contents

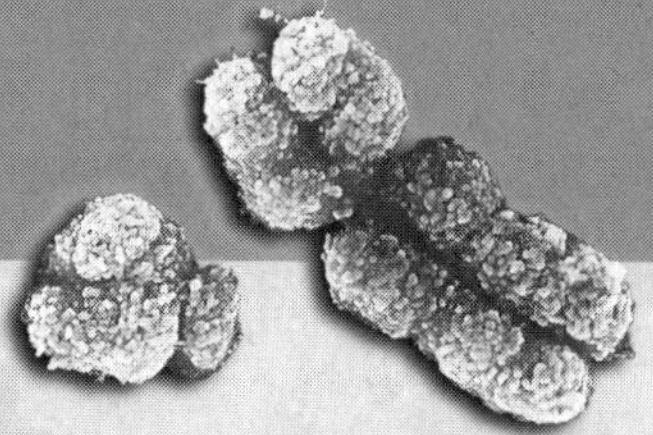

Contents

Hypotheses and Predictions (page 15)

1. Prediction: Woodlice are more likely to be found in moist habitats than in dry habitats.

2. (a) **Bacterial cultures**:
   Prediction: Bacterial strain A will grow more rapidly at 37°C than at room temperature (19°C).
   Outline of the investigation: Set up agar plates of bacterial strain A, using the streak plating method. Place 4 plates in a 37°C incubator and 4 on the lab bench. Leave all 8 plates for the same length of time (e.g. 24 hours), with all other conditions identical. Measure the coverage of the agar plates with bacteria (as a percentage).
   (b) **Plant cloning**:
   Prediction: A greater concentration of hormone A increases the rate of root growth in plant A.
   Outline of the investigation: Set up 6 agar plates infused with increasing concentrations of hormone A (e.g. 1 mgl^{-1}, 5 mgl^{-1}, 10 mgl^{-1}, 50 mgl^{-1}, 100 mgl^{-1}, 500 mgl^{-1}), and each plate with 12 clones of plant A. Measure root length each day for 20 days.

Planning an Investigation (page 17)

1. Aim: To investigate the effect of temperature on the rate of catalase activity.

2. Hypothesis: The rate of catalase activity is dependent on temperature.

3. (a) Independent variable: Temperature.
   (b) Values: 10-60°C in uneven steps: 10°C, 20°C, 30°C, 60°C.
   (c) Unit: °C
   (d) Equipment: A means to maintain the test-tubes at the set temperatures, e.g. water baths. Equilibrate all reactants to the required temperatures in each case, before adding enzyme to the reaction tubes.

4. (a) Dependent variable: Height of oxygen bubbles.
   (b) Unit: mm
   (c) Equipment: Ruler; place vertically alongside the tube and read off the height (directly facing).

5. (a) Each temperature represents a treatment.
   (b) No. of tubes at each temperature = 2
   (c) Sample size: for each treatment = 2
   (d) Times the investigation repeated = 3

6. It would have been desirable to have had an extra tube with no enzyme to determine whether or not any oxygen was produced in the absence of enzyme.

7. Variables that might have been controlled (a-c):
   (a) Catalase from the same batch source and with the same storage history. Likewise for the H_2O_2. Storage and batch history can be determined.
   (b) Equipment of the same type and size (i.e. using test-tubes of the same dimensions, as well as volume). This could be checked before starting.
   (c) Same person doing the measurements of height each time. This should be decided beforehand.

   Note that some variables were controlled: The test-tube volume, and the volume of each reactant. Control of measurement error is probably the most important after these considerations.

8. Controlled variables should be monitored carefully to ensure that the only variable that changes between treatments (apart from the biological response) is the independent (manipulated) variable.

Experimental Method (page 19)

1. Increasing the sample size is the best way to take account of natural variability. In the example described, this would be increasing the number of plants per treatment. **Note**: Repeating the entire experiment as separate trials (as described) is a compromise, usually necessitated by a lack of equipment and other resources. It is not as good as increasing the sample size in one experiment run at the same time, but it is better than just the single run of a small sample size.

2. If all possible variables except the one of interest are kept constant, then you can be more sure that any changes you observe in your experiment (i.e. differences between experimental treatments) are just the result of changes in the variable of interest.

3. Only single plants were grown in each pot to exclude the confounding effects of competition between plants (this would occur if plants were grown together).

4. Physical layout can affect the outcome of experimental treatments, especially those involving growth responses in plants. For example, the physical conditions might vary considerably with different placements along a lab bench (near the window vs central). Arranging treatments to minimise these effects is desirable.

   Checklist to be completed by the student.

Recording Results (page 21)

1. See the results table at the top of the next page.

2. The table would be three times as big in the vertical dimension; the layout of the top of the table would be unchanged. The increased vertical height of the table would accommodate the different ranges of the independent variable (full light, as in question 1, but also half light, and low light. These ranges would have measured (quantified) values attached to them.

Variables and Data (page 22)

1. Measure wavelength (in nm) using a spectrophotometer; which measures light intensity as a function of the colour (wavelength) of light.

2. These data are semi-quantitative because an arbitrary numerical value has been assigned to a qualitative scale. The numbers are correct in a relative sense, but do not necessarily indicate the true quantitative values.

Transforming Raw Data (page 23)

1. (a) Transforming data involves performing calculations using the raw data to determine such properties as rates, percentages, and totals.
   (b) The purpose of data transformation is to convert raw data into a more useful form.

2. (a) **Transformation**: Percentage (percentage cover)
   Reason: Abundance alone might not reflect the importance of a species in terms of its dominance in the habitat.

		Trial 1 / CO_2 conc. in ppm											Trial 2 / CO_2 conc. in ppm											Trial 3 / CO_2 conc. in ppm										
		Minutes											Minutes											Minutes										
	Set up no.	0	1	2	3	4	5	6	7	8	9	10	0	1	2	3	4	5	6	7	8	9	10	0	1	2	3	4	5	6	7	8	9	10
Full light conditions	1																																	
	2																																	
	3																																	
	Av.																																	

(b) **Transformation**: Relative value (ml per unit weight)
Reason: this transformation allows animals of different body size to be compared meaningfully without the interfering influence of actual body size.

(c) **Transformation**: Reciprocal
Reason: Provides a measure of rate where the data have been recorded over very different time periods (time taken for precipitation to occur). It is difficult to compare values where the time scale is different for each recording.

(d) **Transformation**: Rate
Reason: Data may have been recorded over different time periods. A rate allows the production of CO_2 to be compared per unit of time over all temperatures (removes the confounding effect of different time periods as well as different temperatures).

3. Performing data transformations:

(a) Incidence of cyanogenic clover in different regions:

Clover type	Frost free No.	Frost free %	Frost prone No.	Frost prone %	Totals
Cyanogenic	124	78	26	18	150
Acyanogenic	35	22	115	82	150
Total	159	100	141	100	300

(b) Plant transpiration loss:

Time/ *min*	Pipette arm reading/ cm^3	Plant water loss/ $cm^3\ min^{-1}$
0	9.0	-
5	8.0	0.20
10	7.2	0.16
15	6.2	0.20
20	4.9	0.26

(c) Photosynthetic rate at different light intensities:

Light intensity/ %	Average time/ *min*	Reciprocal of time/ min^{-1}
100	15	0.067
50	25	0.040
25	50	0.020
11	93	0.011
6	187	0.005

(d) Frequency of size classes of eels:

Size class/ *mm*	Frequency	Relative frequency/ %
0-50	7	2.6
50-99	23	8.5
100-149	59	21.9
150-199	98	36.3
200-249	50	18.5
250-299	30	11.1
300-349	3	1.1
Total	**270**	**100.0**

Data Presentation (page 26)

1. The difference between the two means (labelled A) is not significant, i.e. the two means are not significantly different because the 95% CIs overlap. The mean at 4 g m^{-3} has such a large 95% CI we cannot be confident that it is significantly different from the mean at 3 g m^{-3} with the very small 95% CI.

2. Graphs and tables provide different ways of presenting information and each performs a different role. Tables summarise raw data, show any data transformations, descriptive statistics, and results of statistical tests. They provide access to an **accurate** record of the data values (raw or calculated), which may not be easily obtained from a graph. Graphs present information in a way that makes any trends or relationships in the data apparent. Both are valuable for different reasons. **Note**: Even when you have calculated descriptive statistics for your data and tabulated these for the reader, it is a good idea to include your raw results as an appendix, or at least have them available for scrutiny.

Drawing Bar Graphs (page 27)

1. (a) Table as below:

Species	**Site 1**	**Site 2**
Ornate limpet	21	30
Radiate limpet	6	34
Limpet sp. A	38	-
Limpet sp. B	57	39
Limpet sp. C	-	2
Catseye	6	2
Topshell	2	4
Chiton	1	3

(b) Bar graph: *See page 4 of graph solutions.*

Drawing Histograms (page 28)

1. (a) Tally chart totals as below:

Weight group	Total
45-49.9	1
50-54.9	2
55-59.9	7 (given)
60-64.9	13
65-69.9	15
70-74.9	13
75-79.9	11
80-84.9	16
85-89.9	9
90-94.9	5
95-99.9	2
100-104.9	0
105-109.9	1

(b) Histogram: *See the next page of graph solutions.*

Drawing Pie Graphs (page 29)

1. (a) Tabulated data:

Food item in diet	Stoats		Rats		Cats	
	% in diet	Angle /°	% in diet	Angle /°	% in diet	Angle /°
Birds	23.6	85	1.4	5	6.9	25
Crickets	15.3	55	23.6	85	-	-
Insects	15.3	55	20.8	75	1.9	7
Voles	9.2	33	-	-	19.4	70
Rabbits	8.3	30	-	-	18.1	65
Rats	6.1	22	-	-	43.1	155
Mice	13.9	50	-	-	10.6	38
Fruits	-	-	40.3	145	-	-
Leaves	-	-	13.9	50	-	-
Unid.	8.3	30	-	-	-	-

(b) Pie graphs: *See the next page of graph solutions.*

Drawing Kite Graphs (page 30)

1. (a) Table:

Distance from mouth/ *km*	Wet weight/ *g* m^{-2}		
	Stm A	Stm B	Stm C
0	0.4	0.4	0
0.5	0.5	0.6	0.5
1.0	0.4	0.1	0
1.5	0.3	0.5	0.2
2.0	0.3	0.4	-
2.5	0.6	0.3	-
3.0	0.1	-	-
3.5	0.7	-	-
4.0	0.2	-	-
4.5	2.5	-	-
5.0	0.3	-	-

(b) Kite graph: *See the next page of graph solutions.*

Drawing Line Graphs (page 31)

1. (a) Line graph:

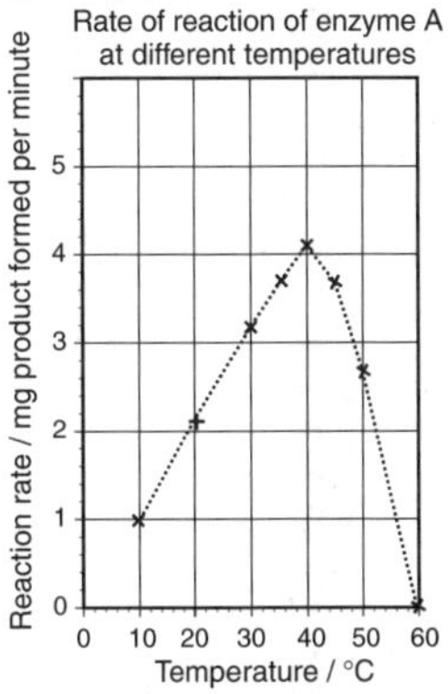

(b) Rate of reaction at 15°C = 1.6 mg product min^{-1}

2. (a) Line graph: See next page of graph solutions.
   (b) The data suggest that the deer population is being controlled by the wolves. Deer numbers increase to a peak when wolf numbers are at their lowest; the deer population then declines (and continues declining) when wolf numbers increase and then peak. **Note**: A scenario of apparent control of the deer population by the wolves is suggested, but not confirmed, by the data. In natural systems, this suggestion (of prey control by a large predator) *may* be specious; most large predators do not control their prey (except perhaps at low population densities in certain systems), but are themselves controlled by the numbers of available prey, which are regulated by other factors such as food availability. In this case, the wolves were introduced for the purpose of controlling deer and were probably doing so. However, an equally valid interpretation of the data could be that the wolves are responding to changes in deer numbers (with the usual lag inherent in population responses), and the deer were already peaking in response to factors about which we have no information.

3. (a) Line graph and (b) point at which shags and nests were removed: See the top of page 5.

Interpreting Line & Scatter Graphs (page 34)

1. (b) **Slope**: Negative linear relationship, with constantly falling slope.
   Interpretation: Variable Y decreases steadily with increase in variable X.
   (c) **Slope**: Constant, level slope.
   Interpretation: Increase in variable X does not affect variable Y.
   (d) **Slope**: Slope rises and then becomes level.
   Interpretation: Variable Y initially increases with increase in variable X, then levels out (no further increase with increase in variable X).
   (e) **Slope**: Rises, peaks and then falls.
   Interpretation: Variable Y initially increases with increase in variable X, peaks and then declines with further increase in variable X.
   (f) **Slope**: Exponentially increasing slope.

Drawing bar graphs:

Average abundance of eight mollusc species at two sites along a rocky shore

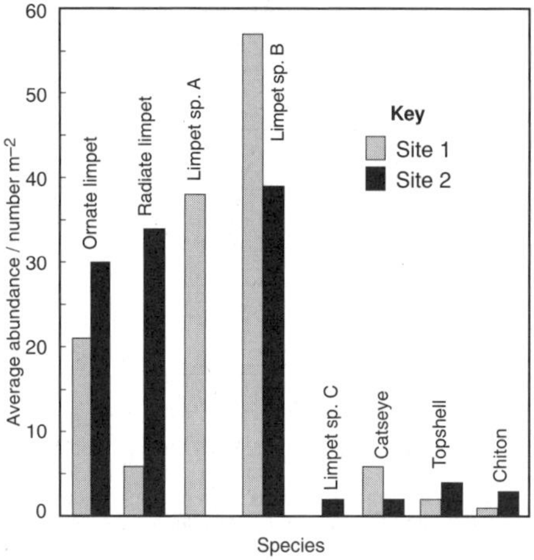

Drawing histograms:

Frequency histogram of weights of 95 individual males and females

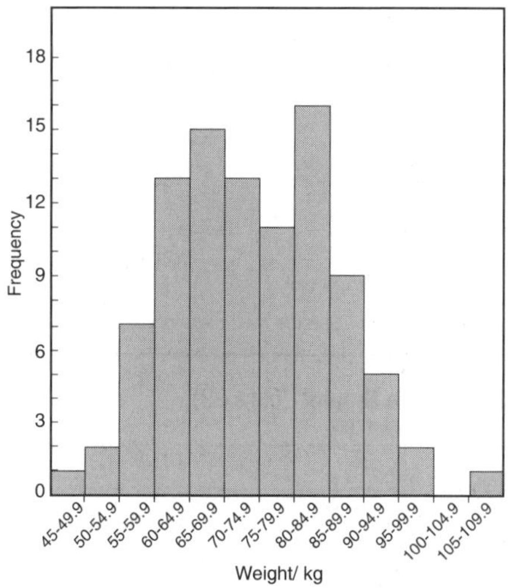

Drawing pie graphs:

Percentage occurrence of different food items in the diets of stoats, rats, and cats

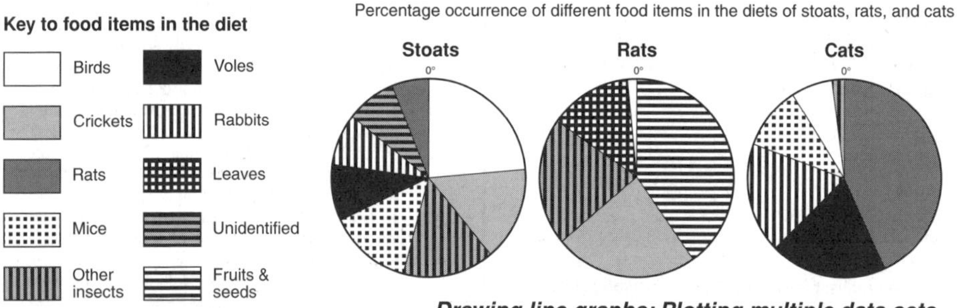

Drawing kite graphs:

Distribution of invertebrates along 3 different streams as indicated by biomass measured as wet weight/ g m^{-2}

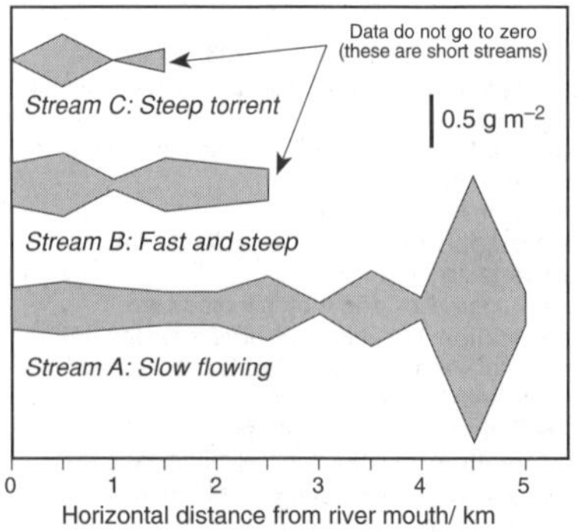

Drawing line graphs: Plotting multiple data sets

Numbers of deer and wolves on an island forest reserve between 1961 and 1969

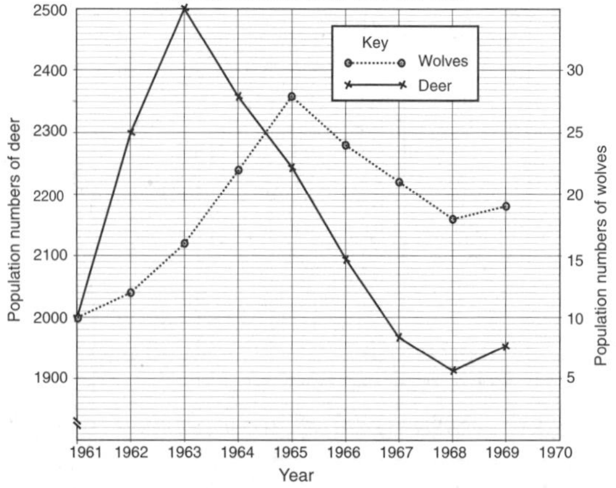

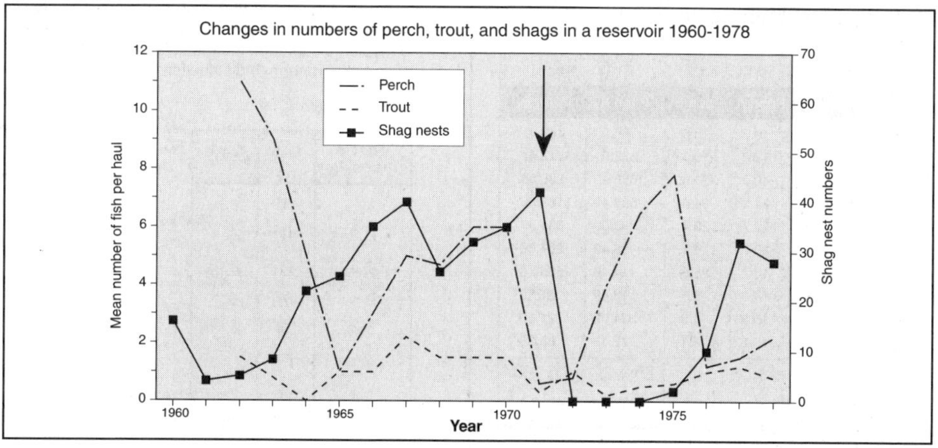

Interpretation: As variable X increases, variable Y increases exponentially.

2. The data suggest that the deer population is being controlled by the wolves. Deer numbers increase to a peak when wolf numbers are at their lowest; the deer population then declines (and continues declining) when wolf numbers increase and then peak. **Note**: A scenario of apparent control of the deer population by the wolves is suggested, but not confirmed, by the data. In natural systems, this may be a specious suggestion; most large predators do not usually control their prey, but are themselves controlled by the numbers of available prey, which are regulated by other factors such as food availability. In this case, the wolves were introduced for the purpose of controlling deer and were probably doing so. However, an equally valid interpretation of the data could be that the wolves are responding to changes in deer numbers (with the usual lag inherent in population responses), and the deer were already peaking in response to factors about which we have no information.

Drawing Scatter Plots (page 35)

1. (a) Scatter plot and (b) Line of best fit:

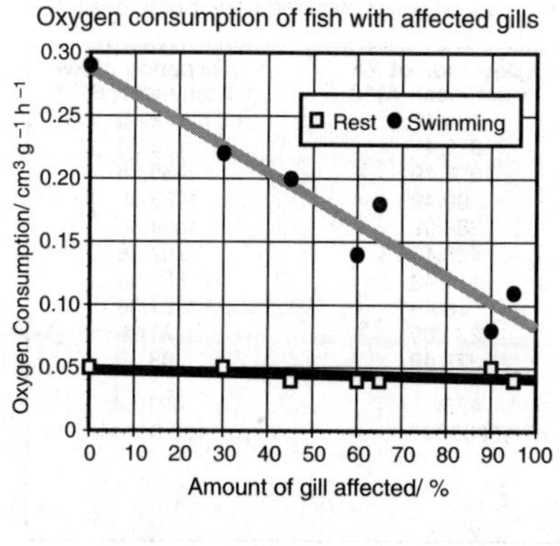

2. (a) At rest: No clear relationship; the line on the graph appears to have no significant slope (although this could be tested). **Note**: There is a slight tendency for oxygen consumption to fall as more of the gill becomes affected, but the scatter of points precludes making any conclusions about this.
   (b) Swimming: A negative linear relationship; the greater the proportion of affected gill, the lower the oxygen consumption.

3. The gill disease appears to have little or no effect on the oxygen uptake in resting fish.

The Student's *t* Test (page 38)

1. (a) The calculated *t* value is less than the critical value of $t = 2.57$. The null hypothesis cannot be rejected. (There is no difference between the control and the experimental treatments).
   (b) The new *t* value supports the alternative hypothesis at $P = 0.05$ (reject the null hypothesis and conclude that there is a difference between the control and experimental treatments). Note the critical value of *t* in this case is 2.23 at 10 d.f. $P = 0.05$

2. Outliers can skew the data set, leading to mean values between data sets that are very different (even though the bulk of the data may not be very different). This may result in inappropriate rejection of the null hypothesis.

3. Statistical significance refers to the probability that an observed difference (or trend) will occur by chance. It is an arbitrary criterion used as the basis for accepting or rejecting the null hypothesis in an investigation. **Note**: In science the term 'significantly different' has a specific meaning. It should not be used in a casual manner when no statistical test has been performed.

Student's *t* Test Exercise (page 39)

1. (a) Completed table:

x (counts)		$x - \bar{x}$		$(x - \bar{x})^2$	
Popn A	Popn B	Popn A	Popn B	Popn A	Popn B
465	310	9.3	-10.6	86.5	112.4
475	310	19.3	-10.6	372.5	112.4
415	290	-40.7	-30.6	1656.49	936.36
480	355	24.3	34.4	590.49	1183.36
436	350	-19.7	29.4	388.09	864.36
435	335	-20.7	14.4	428.49	207.36
445	295	-10.7	-25.6	114.49	655.36
460	315	4.3	-5.6	18.49	31.36
471	316	15.3	-4.6	234.09	21.16
475	330	19.3	9.4	372.49	88.36
$n_A = 10$	$n_B = 10$			$\Sigma(x - \bar{x})^2$	$\Sigma(x - \bar{x})^2$
				4262.1	4212.4

The number of samples in each data set

(b) Variance of population A: 473.57
Variance of population B: 468.04
(c) Difference between population means: 135.1
(d) *t* value = 13.92
(e) Degrees of freedom: 18
(f) $P = 0.05$ *t* (critical value) = 2.101
(g) Decision: We can reject the null hypothesis of no difference. The difference between the population means is significantly different at $P = 0.05$.
Note at $P = 0.001$, the critical *t* value is 3.922, so we can also reject the null hypothesis at $P = 0.001$.

2. (a) See spreadsheet below. Calculated values are:
Mean: A= 455.7, B = 320.6
Difference between means: 135.1
Sum of squares: A= 4262.1, B = 4212.4
Variance: A = 473.567, B = 468.044
$t = 13.92$
(b) New *t* value: 0.76
Decision: We cannot reject the null hypothesis (P= 0.05). There is no difference between population means.

Descriptive Statistics (page 43)

1. The modal value and associated ranked entries indicate that the variable being measured (spores per frond) has a bimodal distribution i.e. the data are not normally distributed. (Therefore) the mean and median are not accurate indicators of central tendency. Note also that the median differs from the mean; also an indication of a skewed (non-normal) distribution.

2. See results below:

Beetle mass /g	Tally	Total
2.1	I	1
2.2	II	2
2.4	II	2
2.5	IIII	4
2.6	III	3
2.7	I	1
2.8	II	2

Median = 8th value when in rank order = 2.5

Mode = 2.5

Mean = 2.49 ~ 2.5

The Reliability of the Mean (page 45)

Although this is an activity, there is no model answer. Students can follow the steps outlined in the worked example and recreate the figures for themselves. The full spreadsheet analysis of this activity is provided on the Teacher Resource CD-ROM.

Biological Drawings (page 47)

1. (a)-(h) any eight features in any order:
   - Lines cross over each other and are angled.
   - Cells are inaccurately drawn: they are not closed shapes, they do not even nearly represent what is actually there, there are overlaps.
   - There is no magnification given.
   - Drawing is cramped at the top corner of the page.
   - Labels are drawn on an angle.
   - There is no indication of whether the section is a cross section or longitudinal section.
   - There is a line to a cell type that has no label
   - Shading is inappropriate and does not indicate anything. It is apparently random and is unnecessary.
   - The material being drawn has not been identified accurately in the title by species.

2. Student's response required here. Some desirable

XA	XB	Deviation of XA from mean A	Deviation of XB from mean B	(Deviation of XA from mean A)^2	(Deviation of XB from mean B)^2
465	310	9.3	-10.6	86.49	112.36
475	310	19.3	-10.6	372.49	112.36
415	290	-40.7	-30.6	1656.49	936.36
480	355	24.3	34.4	590.49	1183.36
436	350	-19.7	29.4	388.09	864.36
435	335	-20.7	14.4	428.49	207.36
445	295	-10.7	-25.6	114.49	655.36
460	315	4.3	-5.6	18.49	31.36
471	316	15.3	-4.6	234.09	21.16
475	330	19.3	9.4	372.49	88.36
Totals 4557	3206		**Sum x^2**	4262.1	4212.4
Count 10	10		**s^2**	473.5666667	468.04444
Mean 455.7	320.6				
Difference between means		135.1			
t value 13.92257486					

features are shown in the figure on the top of the next column, but page position and size cannot be shown.

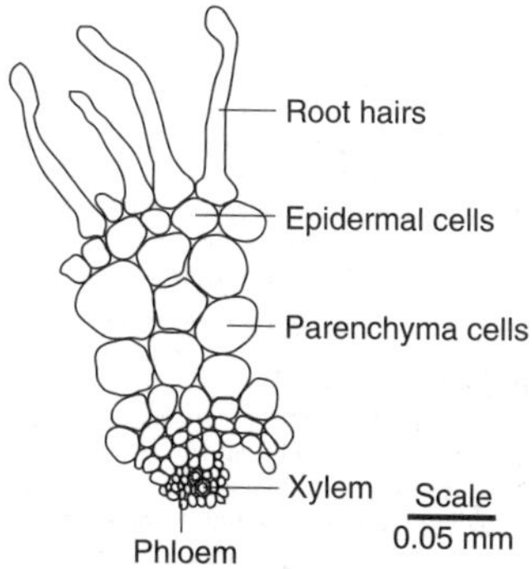

Root tranverse section from Ranunculus

3. A **biological drawing** is designed to convey useful information about the structure of an organism. From such diagrams another person should be able to clearly identify similar organisms and structures. By contrast, **artistic drawings** exhibit 'artistic licence' where the image is a single person's impression of what they saw. It may not be a reliable source of visual information about the structure of the organism.

The Structure of a Report (page 49)

1. (b) **Methods**: provides the reader with instructions on how the investigation was carried out and what equipment was used. Allows for the procedures to be repeated and confirmed by other investigators.
   (c) **Results:** Provides the findings of the investigation and allows the reader to evaluate these themselves.
   (d) **Discussion:** The findings of the work are discussed in detail so the reader can evaluate the findings. Design limitations, and ways the work could have been improved are also presented.
   (e) **References/Acknowledgments:** Lists sources of information and help used during the investigation. The reader can review the references for more detail if required, and compare your work with other studies in the area of investigation.

2. A poster presents all of the key information from an investigation in an attractive, concise manner which is readily accessible and easy to read. People can quickly determine if the study is of interest to them, and the references provide an opportunity to find out further information if required.

Writing the Methods (page 50)

1. (a)-(h) Any of the following in any order:
   - Number of worms used not stated.
   - No description of the pond (size, water depth etc.).
   - Actual "room" temperature not stated.
   - Date somewhat irrelevant (time of year could be).
   - Source of sea water not stated.
   - Pre-experimental conditions of the worms not stated.
   - Volume of 100% sea water used not stated.
   - Dilution of sea water not stated.
   - Volume of diluted sea water used not stated.
   - Weighing equipment used not stated.
   - Time interval for reweighing not stated.

Writing Your Results (page 51)

1. Referring to tables and figures in the text clearly indicates which data you are referring to in your synopsis of the results and gives the reader access to these data so that they can assess your interpretation.

2. Tables summarise data and provide a record of the data values, which may not be easily obtained from a graph. Figures (graphs) present information in a way that makes trends or relationships in the data apparent. Such trends may not be evident from the tabulated data. Both formats are valuable for different reasons.

Writing Your Discussion (page 52)

1. Discussion of weaknesses in your study shows that you have considered these and acknowledged them and the effect that they may have had on the outcome of your investigation. It also provides the opportunity for those repeating the investigation (including yourself) to improve on aspects of the design.

2. A **critical evaluation** shows that you have examined your results carefully in light of the question(s) you asked and the predictions you made. Objective evaluation enables you to provide reasonable explanations for any unexpected or conflicting results and identify ways in which to improve your study design in future investigations.

3. The conclusion allows you to make a clear statement about your findings, i.e. whether for not the results support your hypothesis. If your results and discussion have been convincing, the reader should be in agreement with the conclusion you make.

Citing and Listing References (page 53)

1. A bibliography lists all sources of information whereas a reference list includes only those sources that are cited in the text. Usually a bibliography is used to compile the final reference list, which appears in the report.

2. Internet articles can be updated as new information becomes available and the original account is revised. It is important that this is noted because people using that source in the future may find information that was unavailable to the author making the original citation.

3. Reference list as follows:
   Ball, P. (1996): Living factories. New Scientist, 2015, 28-31
   Campbell, N. (1993): Biology. Benjamin/Cummings. Ca.
   Cooper, G. (1997): The cell: a molecular approach. ASM Press, Washington DC. pp. 75-85
   Moore, P. (1996): Fuelled for Life. New Scientist, 2012, 1-4
   O'Hare, L. & O'Hare, K. (1996): Food biotechnology. Biological Sciences Review, 8(3), 25.
   Roberts, I. & Taylor, S. (1996): Development of a procedure for purification of a recombinant therapeutic protein. Australasian Biotechnology, 6(2), 93-99.

Report Checklist (page 55)

To be competed by the student.

The Cell Theory (page 57)

1. Microscopes enabled us to see and examine cells in detail. Microscopy opened up an entire new field: the study of cells and microorganisms.

2. Spontaneous generation referred to the arising of living matter from non-living (inanimate) material (e.g. blowflies arising from meat). It was discredited because closer examination of cells and their processes revealed how cells really arise, grow, and divide.

Characteristics of Life (page 58)

1. Cytoplasm (nutrient "soup"), plasma membrane, metabolism (the cell's own cellular machinery).

2. (a) **Size**: Viruses are very small: generally 50-500 times smaller than a typical prokaryotic cell and up to 5000 times smaller than a eukaryote cell.
   (b) **Metabolism**: Cells have metabolic activity; there are chemical reactions taking place much of the time. A virus has no cytoplasm and no metabolism of its own. It relies on the metabolism of its host cell.
   (c) **Organelles**: Viruses have no organelles unlike cells, most of which have organelles which carry out specific roles in the cell.
   (d) **Genetic material**: Viruses have a single or double stranded chromosome which can be RNA or DNA. Cells have only double stranded DNA chromosomes. In eukaryotes the chromosomes are contained within a nuclear membrane.
   (e) **Life cycle**: Outside a living cell viruses exist as inert particles, adopting a "living" programme only when they invade a host cell and can take over the cellular machinery of the cell. At times, they may integrate into the host cell's chromosome and remain latent. Cells are generally either "alive" (when there is metabolic activity) or dead (no metabolic activity). **Note**: There are exceptions to this generalisation, e.g. bacterial endospores, which are special resting stages with no metabolic activity.

3. Multicellular organisms are said to show emergent properties because they have properties that go beyond those of a single cell or unicellular organism, e.g. specialised cell types with differing functions.

Types of Living Things (page 59)

1. (a) **Autotrophic**: Plant cells, some protoctistan cells, some bacterial cells.
   (b) **Heterotrophic**: Fungal cells, animal cells, some protoctistan cells, some bacterial cells, viruses.

2. (a) Prokaryotic cells are much smaller (and simpler) than the cells of eukaryotes.
   (b) Prokaryotic cells are bacterial cells while eukaryotic cells are all cell types other than bacteria and viruses. **Note**: More specifically, prokaryotes lack a distinct nucleus, have no membrane-bound organelles, have a cell wall usually containing peptidoglycan and their DNA is present as a single, naked chromosome.

3. (a) Fungi are plant-like in their appearance and habit (growth form, lack of movement, habitat etc.).
   (b) This classification was erroneous because fungi (unlike plants) lack chlorophyll, their cell walls contain chitin (not cellulose), and they are heterotrophic (not autotrophic).

4. Protoctistans often exhibit both animal-like and plant-like features and the group is very diverse in terms of nutrition, reproduction, and structure. **Note**: From a phylogenetic point of view, the protoctists are not monophyletic and should be classified accordingly.

Types of Cells (page 60)

1. (a) **Plant cells**: Mesophyll cell, vessel element (note that this is a dead cell), guard cells.
   (b) **Characteristics**: Cellulose cell wall, chloroplasts containing the photosynthetic pigments chlorophyll *a* and *b*, carbohydrate stored as starch, large vacuoles.

2. (a) **Animal cells**: Osteocyte, leucocyte, smooth muscle cell, epidermal cells of skin, neurone, erythrocyte.
   (b) **Characteristics**: No cellulose cell wall, no chloroplasts or plastids of any kind, vacuoles if present are small, no regular geometric shape.

3. (a) **Protoctistans**: *Amoeba, Euglena, Paramecium, Spirogyra*.
   (b) **Characteristics**: A very diverse group (the eukaryotes that do not fit into plant, animal, or fungi classification). Generally refers to unicellular eukaryotes, although some primitive multicellular organisms have been included in recent years. Includes protozoans which are heterotrophic (animal-like in their nutrition), and algae, which have chloroplasts and are autotrophic (plant-like in their nutrition). A difficult group to give general characteristics for because they are so diverse. Often ciliated (e.g. *Paramecium*) or flagellated (e.g. *Euglena*). Note that the cyanobacteria are now classified within Prokaryotae (not Protoctista).

4. (a) **Fungal cells**: Yeast, pin mould hyphae.
   (b) **Characteristics**: Eukaryotes, usually multicellular, all are heterotrophs (no photosynthetic ability), lack chloroplasts. Except in yeasts, the basic fungus body is composed of basic building blocks called hyphae. Cell walls composed mainly of chitin.

5. (a) **Prokaryotae**: *Cyanobacteria*, *Diplococcus* bacteria, *Bacillus* bacteria.
   (b) **Characteristics**: Prokaryotes (single, circular chromosome not contained within a nuclear membrane), cell wall containing petidoglycan material (not cellulose), extra-chromosomal material may be present (plasmids), some bacteria capable of forming resistant endospores.

Cell Sizes (page 61)

1.

(a) *Amoeba*:	300 µm	0.3 mm
(b) Foraminiferan:	400 µm	0.4 mm
(c) *Leptospira*:	7-8 µm	0.007-0.008 mm
(d) Epidermis:	120 µm	0.12 mm
(e) *Daphnia*:	2500 µm	2.5 mm
(f) *Papillomavirus*:	0.13 µm	0.00013 mm

2. *Papillomavirus*; *Leptospira*; Epidermis (but not an organism); *Amoeba*; Foraminiferan; *Daphnia*

3. Onion epidermis (possibly); *Amoeba*; foraminiferan; *Daphnia*

4. (a) 0.00025 mm (b) 0.45 mm (c) 0.0002 mm

Plant Cells (page 62)

1. A: Nucleus B: Cell wall
   C: Nucleus D: Chloroplasts

2. (a) Cytoplasmic streaming is the rapid movement of cytoplasm within eukaryotic cells, seen most clearly in plant and algal cells.
   (b)

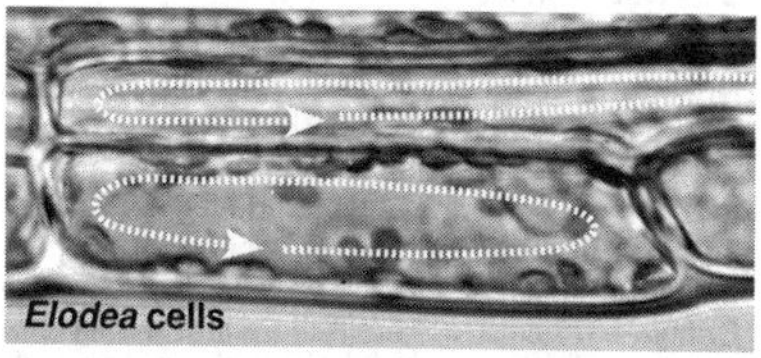

3. Any three of:
   - Starch (branched carbohydrate) granules stored in amyloplasts (energy store).
   - Chloroplasts, discrete plastids containing the pigment chlorophyll, involved in photosynthesis.
   - Large vacuole, often central (vacuoles are present in animal cells, but are only small).
   - Cell wall of cellulose forming the rigid, supporting structure outside the plasma membrane.
   - Plasmodesmata.

Animal Cells (page 63)

1. A: Nucleus
   B: Plasma membrane
   C: Nucleus

2. (a)

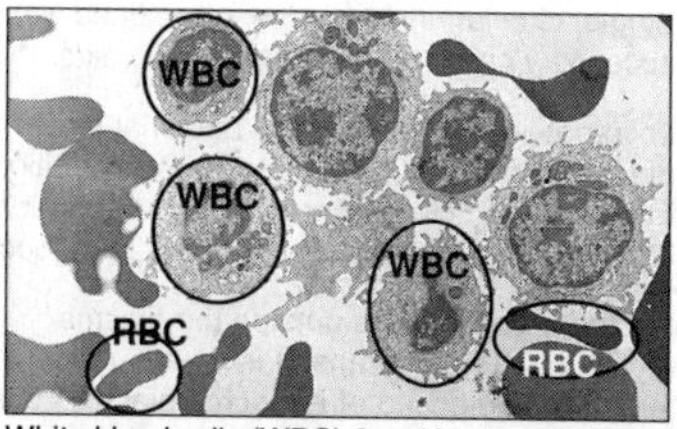

White blood cells (WBC) & red blood cells (RBC)

   (b) Any of the following: RBCs have no nucleus and they are smaller than the white blood cells. White blood cells have extensions of the plasma membrane (associated with being phagocytic ad mobile), are larger than RBCs, and have a nucleus.

3. Any one of:
   – Centrioles (although these are present in lower plants, they are absent from higher plants). They are microtubular structures responsible for forming the poles and the spindles during cell division.
   – Desmosomes. These are points of contact between the plasma membranes of neighbouring cells, which allow cells to combine together to form tissues.

Cell Structures and Organelles (page 64)

(b) **Name**: Ribosome
Location: Free in cytoplasm or bound to rough ER
Function: Synthesize polypeptides (=proteins)
Present in plant cells: Yes
Present in animal cells: Yes
Visible under LM: No

(c) **Name**: Mitochondrion
Location: In cytoplasm as discrete organelles
Function: Site of cellular respiration (ATP formation)
Present in plant cells: Yes
Present in animal cells: Yes
Visible under LM: Not with most standard school LM, but can be seen using high quality, high power LM.

(d) **Name**: Golgi apparatus
Location: In cytoplasm associated with the smooth endoplasmic reticulum, often close to the nucleus.
Function: Final modification of proteins and lipids. Sorting and storage for use in the cell or packaging molecules for export.
Present in plant cells: Yes
Present in animal cells: Yes
Visible under LM: Not with most standard school LM, but may be visible using high quality, high power LM.

(e) **Name**: Endoplasmic reticulum (in this case, rough ER)
Location: Penetrates the whole cytoplasm
Function: Involved in the transport of materials (e.g. proteins) within the cell and between the cell and its surroundings.
Present in plant cells: Yes
Present in animal cells: Yes
Visible under LM: No

(f) **Name**: Chloroplast
Location: Within the cytoplasm
Function: The site of photosynthesis
Present in plant cells: Yes
Present in animal cells: No
Visible under LM: Yes

(g) **Name**: Lysosome and food vaculoe (given)
Lysosome
Location: Free in cytoplasm.
Function: Ingests and destroys foreign material. Able to digest the cell itself under some circumstances.
Present in plant cells: Yes but variably (vacuoles may have a lysosomal function in some plant cells).
Present in animal cells: Yes
Visible under LM: No
Vacuole (a food vacuole in an animal cell is shown, so students may answer with respect to this).
Location: In cytoplasm.
Function: In plant cells, the vacuole (often only one) is a large fluid filled structure involved in storage and support (turgor). In animal cells, vacuoles are smaller and more numerous, and are involved in storage (of water, wastes, and soluble pigments).
Present in plant cells: Yes, as (a) large structure(s).
Present in animal cells: Yes, smaller, more numerous
Visible under LM: Yes in plant cells, no in animal cells.

(h) **Name**: Nucleus
Location: Discrete organelle, position is variable.
Function: The control center of the cell; the site of the nuclear material (DNA).
Present in plant cells: Yes

Present in animal cells: Yes
Visible under LM: Yes.

(i) **Name**: Centrioles
Location: In cytoplasm, usually next to the nucleus.
Function: Involved in cell division (probably in the organisation of the spindle fibers).
Present in plant cells: Variably (absent in higher plants)
Present in animal cells: Yes
Visible under LM: No.

(j) **Name**: Cilia and flagella (given)
Location: Anchored in the cell membrane and extending outside the cell.
Function: Motility.
Present in plant cells: No
Present in animal cells: Yes
Visible under LM: Variably (depends on magnification and preparation/fixation of material).

(k) **Name**: Cytoskeleton
Location: Throughout cytoplasm
Function: Provides structure and shape to a cell, responsible for cell movement (e.g. during muscle contraction), and provides intracellular transport of organelles and other structures.
Present in plant cells: Yes
Present in animal cells: Yes
Visible under LM: No

(l) **Name**: Cellulose cell wall
Location: Surrounds the cell and lies outside the plasma membrane.
Function: Provides rigidity and strength, and supports the cell against changes in turgor.
Present in plant cells: Yes
Present in animal cells: No
Visible under LM: Yes

(m) **Name**: Cell junctions (an animal example is given)
Location: At cell membrane surface, connecting adjacent cells.
Function: Depends on junction type. Desmosomes fasten cells together, gap junctions act as communication channels between cells, and tight junctions prevent leakage of extracellular fluid from layers of epithelial cells.
Present in plant cells: Yes, as plasmodesmata
Present in animal cells: Yes
Visible under LM: No

Differential Centrifugation (page 67)

1. Cell organelles have different densities and spin down at different rates. Smaller organelles take longer to spin down and require a higher centrifugation speed to separate out.

2. The sample is homogenized (broken up) before centrifugation to rupture the cell surface membrane, break open the cell, and release the cell contents.

3. (a) Isotonic solution is needed so that there are no volume changes in the organelles.
(b) Cool solution prevent self digestion of the organelles by enzymes released during homogenization.
(c) Buffered solution prevents pH changes that might denature enzymes and other proteins.

4. (a) Ribosomes and endoplasmic reticulum
(b) Lysosomes and mitochondria
(c) Nuclei

Unicellular Eukaryotes (page 68)

1. Summary for each organism under the given headings:

Amoeba:
Nutrition: Heterotrophic, food (e.g. bacteria) ingested by phagocytosis. Food digested in food vacuoles.
Movement: By pseudopodia (cytoplasmic projections).
Osmoregulation: Contractile vacuole.
Eye spot: Absent.
Cell wall: Absent.

Paramecium:
Nutrition: Heterotrophic (feeds on bacteria and small protoctists). Food digested in food vacuoles.
Movement: By beating of cilia.
Osmoregulation: Contractile vacuoles.
Eye spot: Absent.
Cell wall: Absent.

Euglena:
Nutrition: Autotrophic (photosynthetic), but can be heterotrophic when light deprived.
Movement: By flagella (one larger, which is labelled, and one very small, beside the gullet).
Osmoregulation: Contractile vacuole.
Eye spot: Present.
Cell wall: Absent, although there is a wall-like pellicle, which lies inside the plasma membrane and is flexible.

Chlamydomonas:
Nutrition: Autotrophic (photosynthetic).
Movement: By flagella.
Osmoregulation: Contractile vacuole.
Eye spot: Present.
Cell wall: Present.

2. *Amoeba*, *Paramecium*, *Euglena*, *Chlamydomonas.*

3. An eye spot enables an autotroph to detect light and so move into well lit regions where it can photosynthesise.

Prokaryotic Cells (page 69)

1. (a) The nuclear material (DNA) is not contained within a clearly defined nucleus with a nuclear membrane.
(b) Membrane-bound cellular organelles (e.g. mitochondria, endoplasmic reticulum) are missing.
(c) Single, circular chromosome sometimes with accessory chromosomes called plasmids.

2. (a) Locomotion: Flagella enable bacterial movement out of unsuitable conditions to preferred conditions.
(b) Fimbriae are shorter, straighter, and thinner than flagella. Used for attachment rather than locomotion.

3. (a) Bacterial cell wall lies outside the plasma (cell surface) membrane. It is a semi-rigid structure composed of a macromolecule called peptidoglycan, and contains varying amounts of lipopolysacchardies and lipoproteins.
(b) The glycocalyx is a viscous, gelatinous layer which lies outside the cell wall. It usually comprises polysaccharide and/or polypeptide, but not peptidoglycan, and may be firmly or loosely attached to the wall.

4. (a) Bacteria usually reproduce by binary fission, where the DNA replicates and the cell then splits into two.
(b) Conjugation differs from binary fission in that DNA is exchanged between one bacterial cell (the donor) and another (the recipient). The recipient cell gains DNA from the donor.

5. Plasmids are used extensively in recombinant DNA technology. Being accessory to the main chromosome, the plasmid DNA can be manipulated easily. Using restriction enzymes, foreign genes (e.g. gene for producing insulin) can be spliced into a plasmid, which then carries out the instructions of the foreign gene.

Production and Secretion of Proteins (page 71)

1. (a) **Glycoproteins** are proteins with attached carbohydrates (often relatively small polymers of sugar units).
   (b) In any order, three roles of glycoproteins:
   - **Intercellular recognition**: Present on cell surfaces for recognition between cells (when cells interact to form tissues and for immune function).
   - **Transport**: Embedded in cell membranes to transport molecules through the membrane (the sugars help to maintain the position of the glycoprotein in the membrane).
   - **Regulation**: Secretory proteins from glands with a role in regulation, e.g. many pituitary hormones.
2. (a) **Lipoproteins** are proteins with attached fatty acids.
   (b) Lipoproteins transport lipid molecules in the plasma between different organs in the body.
3. Proteins made on free ribosomes are released directly into the cytoplasm; there is no facility for attachment of carbohydrate as this generally requires a packaging region (the Golgi apparatus).
4. Polypeptides are synthesised by membrane-bound ribosomes so that they can be easily threaded through the ER membrane into the cisternal space of the ER. Here they are in place for subsequent modification, packaging, and export.
5. The carbohydrates attached to glycoproteins aid in the recognition or function of the protein so that it is transported to the correct destination and performs its appropriate functional role.
6. (a) Rough ER: Ribosomes on the rough ER assemble the proteins destined for secretion.
   (b) Smooth ER: Synthesis of lipids, e.g. steroid hormones and phospholipids, and packages them into transport vesicles.
   (c) Golgi apparatus: Receives transport vesicles. Modifies, stores, and transports molecules for export around or from the cell.
   (d) Transport vesicles: These bud off the ER and move substances to the Golgi apparatus.
7. Protein orientation in the membrane is important because it is usually critical to the functional role of the protein, e.g. in intercellular recognition or transport.

Optical Microscopes (page 73)

1. (a) Eyepiece lens
   (b) Arm
   (c) Coarse focus knob
   (d) Fine focus knob
   (e) Objective lens
   (f) Mechanical stage
   (g) Condenser
   (h) In-built light source
   (i) Eyepiece lens
   (j) Eyepiece focus
   (k) Focus knob
   (l) Objective lens
   (m) Stage
2. Phase contrast: Used where the specimen is transparent (to increase contrast between transparent structures). **Note**: It is superior to dark field because a better image of the interior of specimens is obtained.
3. (a) Plant cell, any two of: Cell wall, nucleus (may see chromatin if stained appropriately), vacuole, cell membrane (high magnification), Golgi apparatus, mitochondria (high magnification), chloroplast, cytoplasm (if stained), nuclear envelope (maybe).
   (b) Animal cell, any two of: Nucleus (may see chromatin if stained appropriately), centriole, cell membrane (high magnification), Golgi apparatus, mitochondria (high magnification), cytoplasm (if stained).
4. Any of: Ribosomes, microtubules, endoplasmic reticulum, Golgi vesicles (free), nuclear envelope as two layers, lysosomes (animal cells). Also detail of organelles such as mitochondria and chloroplasts.
5. (a) Leishman's stain
   (b) Schultz's solution/iodine solution
   (c) Schultz's solution
   (d) Aniline sulfate/ Schultz's solution
   (e) Methylene blue
   (f) Schultz's solution
6. (a) 600X magnification (b) 600X magnification
7. Bright field, compound light microscopes produce a flat (2-dimensional) image from a thin, transparent sample. Dissecting microscopes produce a 3-dimensional image, revealing the surface details of the specimen.
8. Magnification is the number of times larger an image is than the specimen. Resolution is the degree of achievable detail. The limit of resolution is the minimum distance by which two points in a specimen can be separated and still be distinguished as separate points. **Note**: By adding stronger, or more, lenses, a LM can magnify an image many 1000s of times but its resolution is limited. EMs have greater resolving power because of the very short wavelength of the electrons.

Electron Microscopes (page 75)

1. The limit of resolution (see #8 above) is related to wavelength (about 0.45X the wavelength). The shortest visible light has a wavelength of about 450 nm giving a resolution of 0.45 x 450 nm; close to 200 nm. Points less than 200 nm apart will be perceived as one point or a blur. Electron beams have a shorter wavelength than light so the resolution is much greater (points 0.5 nm apart can be distinguished as separate points; a resolving power that is 400X that of a light microscope).
2. (a) **TEM**: Used to (any of): show cell ultrastructure i.e. organelles; to investigate changes in the number, size, shape, or condition of cells and organelles i.e. demonstrate cellular processes or activities; to detect the presence of viruses in cells.
   (b) **SEM**: Used to (any of): show the surface features of cells, e.g. guard cell surrounding a stoma; to show the surface features of organisms for identification (often used for invertebrates and viruses); for general identification by surface feature, e.g. for pollen used in palaeoclimate or forensic research.
   (c) **Bright field**: Used for (any of): examining prepared sections of tissue for cellular detail; for examining living tissue for large scale movements, e.g. blood flow in capillaries or cytoplasmic streaming.

(d) **Dissecting**: Used for (any of): examining living specimens for surface detail and structures; sorting material from samples (e.g. leaf litter or stream invertebrates; dissecting a small organism where greater resolution than the naked eye is required.

3. A TEM
   B Bright field LM
   C TEM
   D Bright field LM
   E SEM
   F Bright field LM
   G Dissecting LM
   H SEM

Interpreting Electron Micrographs (page 77)

1. (a) Chloroplast
   (b) Plant cells, particularly in leaf and green stems.
   (c) Function: Site of photosynthesis. Captures solar energy to build glucose from CO_2 and water.
   (d)

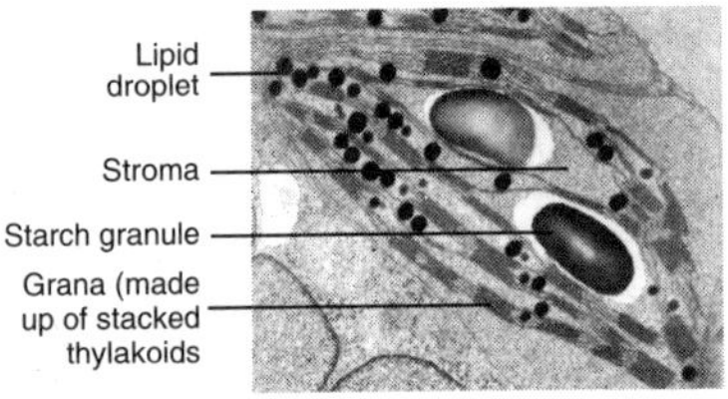

2. (a) Golgi apparatus
   (b) Plant and animal cells
   (c) Function: Packages substances to be secreted by the cell. Forms a membrane vesicle containing the chemicals for export from the cell (e.g. nerve cells export neurotransmitters; endocrine glands export hormones; digestive gland cells export enzymes).

3. (a) Mitochondrion
   (b) Plant and animal cells (most common in cells that have high energy demands, such as muscle).
   (c) Function: Site of most of the process of cellular respiration, which releases energy from food (glucose) to fuel metabolism.
   (d)

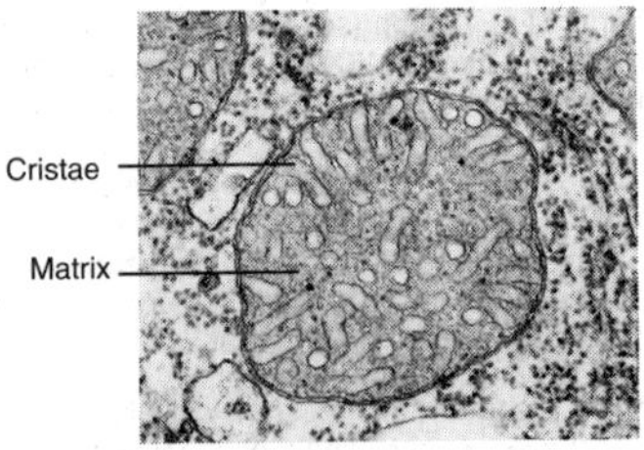

4. (a) Endoplasmic reticulum
   (b) Plant and animal cells (eukaryotes)
   (c) Function: Site of protein synthesis (translation stage). Transport network that moves substances through its system of tubes. Many complex reactions need to take place on the surface of the membranes.
   (d) Ribosomes.

5. (a) Nucleus
   (b) Plant and animal cells (eukaryotes)
   (c) Function: Controls cell metabolism (all the life-giving chemical reactions), and functioning of the whole organism. These instructions are inherited from one generation to the next.
   (d)

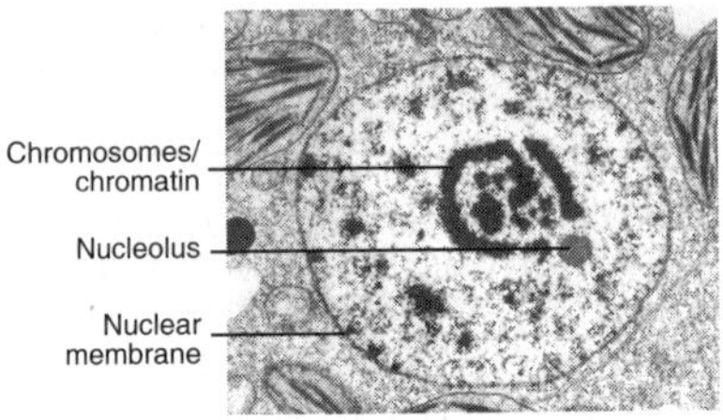

6. (a) Function: Controls the entry and exit of substances into and out of the cell. Maintains a constant internal environment.
   (b)

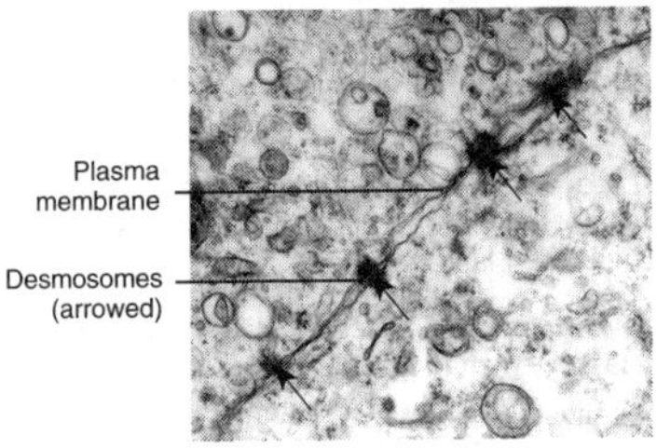

7. Generalised cell.

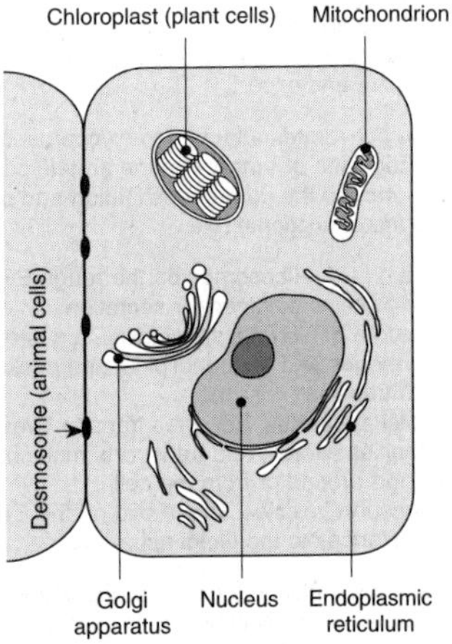

The Role of Membranes in Cells (page 80)

1. (a) Compartments within cells allow specific metabolic pathways in the cell to be localised. This achieves greater efficiency of cell function and restricts potentially harmful reactions and substances (e.g. hydrogen peroxide) to specific areas.
   (b) Greater membrane surface area provides a greater area over which membrane-bound reactions can occur. This increases the speed and efficiency with which metabolic reactions can take place.

The Structure of Membranes (page 81)

1. (a) Membranes are composed of a phospholipid bilayer in which are embedded proteins, glycoproteins, and glycolipids. The structure is relatively fluid and the proteins are able to move within this fluid matrix.
   (b) The Davson-Danielli model described membranes as a lipid bilayer with a coating of protein. This model was modified when freeze-fracture techniques showed that the proteins were embedded in the membrane rather than coating the outside. As described in the fluid-mosaic model, some proteins span the width of the membrane, some are on the outside or the inside.

2. Membranes perform numerous diverse roles. The plasma membrane forms the outer limit of the cell and contains the proteins that confer cellular recognition. It also controls the entry and exit of materials into and out of the cell. Intracellular membranes keep the cytoplasm separate from the extracellular spaces and provide compartments within cells for localisation of metabolic (enzymatic) reactions. They also provide a surface for the attachment of the enzymes involved in metabolism.

3. (a) Any of: Golgi apparatus, mitochondria, chloroplasts, endoplasmic reticulum (rough or smooth), nucleus, vacuoles, lysosomes.
   (b) Depends on choice: Generally the membrane's purpose is to compartmentalise the location of enzymatic reactions, to control the entry and exit of substances that the organelle operates on, and/or to provide a surface for enzyme attachment.

4. Any three of the following not already chosen: Golgi apparatus, mitochondria, vacuoles, endoplasmic reticulum, chloroplasts, lysosomes.

5. (a) Cholesterol lies between the phospholipids and prevents close packing. It thus functions to keep membranes more fluid. The greater the amount of cholesterol in the membrane the greater its fluidity.
   (b) At temperatures close to freezing, high proportions of membrane cholesterol is important in keeping membranes fluid and functioning.

6. (a)-(c) in any order: Oxygen, food (glucose), minerals and trace elements, water.

7. (a) Carbon dioxide (b) Nitrogenous wastes

8.

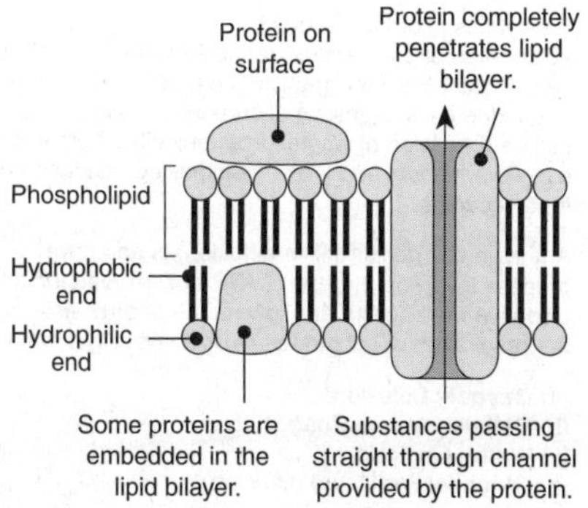

Cell Signalling and Receptors (page 83)

1. Cell signalling is a complex communication system by which cells gather information (chemical signals) about their environment, and change their cellular activity accordingly to respond to it. Cell signalling is also enables communication between cells.

2. (a) Endocrine signalling (b) Paracrine signalling

3. All cell signalling mechanisms have in common some kind of chemical messenger or signal molecule (ligand) and a receptor molecule (on the target cells, which may or may not be on the cell producing the signal).

4. (a) Hormone binding receptors initiate a series of reactions within the cell that result in a physiological change, often to maintain homeostasis. In many cases hormones exert their effects by influencing genes directly or the enzymes associated with membranes. For example, when the hormone insulin binds to a membrane receptor, the cell's glucose transporters open and glucose is taken up from the blood and used for metabolism or storage.
   (b) The binding of a drug to a membrane receptor initiates a physiological response. Two responses by the cell are common. Agonistic drugs cause a stimulatory effect on the cell, while antagonistic drugs block or depress aspects of the cell's activity.

Diffusion (page 84)

1. (a) Large surface area (b) Thin membrane

2. Concentration gradients are maintained by (any one of): Constant use or transport away of a substance on one side of a membrane (e.g. use of ADP in mitochondria). Production of a substance on one side of a membrane (e.g. production of CO_2 by respiring cells).

3. Ionophores allow the preferential passage of some molecules but not others.

Osmosis and Water Potential (page 85)

1. Zero

2. and 3. (a)-(c):

(a)

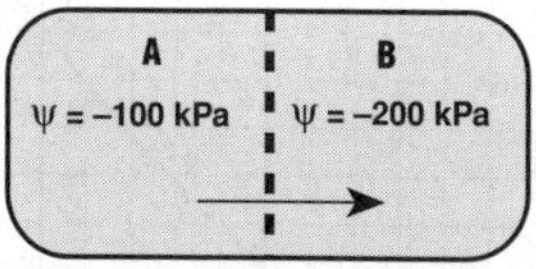

(b)

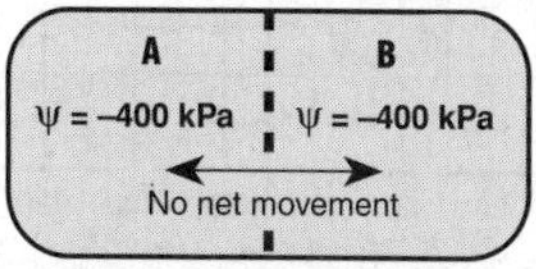

(c)

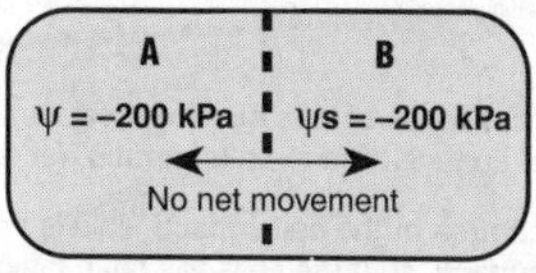

4. (a) Hypotonic
   (b) Fluid replacements must induce the movement of water into the cells and tissues (which are dehydrated and therefore have a more negative water potential than the drink). **Note**: Many sports drinks are isotonic. Depending on the level of dehydration involved, these drinks are more effective when diluted.

5. *Paramecium* is hypertonic to the surrounding freshwater environment and water constantly flows into the cell. This must be continually pumped out (by contractile vacuoles).

6. (a) Pressure potential generated within plant cells provides the turgor necessary for keeping unlignified plant tissues supported.
   (b) Without cell turgor, soft plant tissues (soft stems and flower parts for example) would lose support and wilt. Note that some tissues are supported by structural components such as lignin.

7. Animal cells are less robust than plant cells against changes in net water content: Excess influx will cause bursting and excess loss causes crenulation.

8. (a) Water will move into the cell and it will burst (lyse).
   (b) The cell would lose water and the plasma membrane would crinkle up (crenulate).
   (c) Water will move into the cell and it will burst (lyse).

9. Malarial parasite: **Isotonic** to blood.

Surface Area and Volume (page 87)

1.

Cube	Surface Area	Volume	Ratio
3 cm:	3 x 3 x 6 = 54	3 x 3 x 3 = 27	2.0 to 1
4 cm:	4 x 4 x 6 = 96	4 x 4 x 4 = 64	1.5 to 1
5 cm:	5 x 5 x 6 = 150	5 x 5 x 5 = 125	1.2 to 1

2. Surface area to volume graph: see the next column:

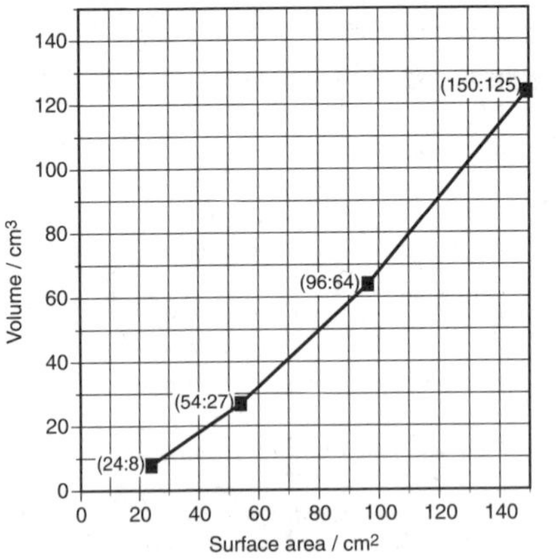

3. Volume

4. Increasing size leads to less surface area for a given volume. The surface area to volume ratio decreases.

5. Less surface area at the cell surface. This is the gas exchange surface, so large cells will have difficulty moving enough materials in and out of the cell to meet demands. This is what limits a cell's maximum size. **Note**: Eukaryote cells are typically about 0.01-0.1 mm in size, but some can be bigger than 1 mm. The largest cell is the female sex cell (ovum) of the ostrich, which averages 15-20 cm in length. Technically a single cell, it is atypical in size because almost the entire mass of the egg is food reserve in the form of yolk, which is not part of the functioning structure of the cell itself.

Ion Pumps (page 89)

1. If an animal cell (not protected by a rigid cell wall), contains excessive quantities of ions, there is a danger that it may take up so much water by osmosis that it would swell up and burst.

2. An ion exchange pump creates an unequal balance of ions across the membrane. The transport of other molecules (e.g. sucrose) can be coupled to the passive diffusion of an ion (e.g. H^+) as it diffuses down its concentration gradient.

3. ATP is required to move ions **against** their concentration gradient (an energy requiring process). (When a phosphate is transferred from the ATP to the carrier protein, a shape change in the protein brings about the transfer of the bound molecule (e.g. an ion) from one side of the membrane to the other).

4. Coupled pumps operate in (any one of):
   - Loading of sucrose into the **phloem sieve tube cells** (coupled to a proton pump).
   - Transport of glucose across the **epithelium** of the gut into the blood (coupled to a sodium pump).

Exocytosis and Endocytosis (page 90)

1. **Phagocytosis** is the engulfment of solid material by endocytosis whereas **pinocytosis** is the uptake of liquids or fine suspensions by endocytosis.

2. Phagocytosis examples (any of):
   • Feeding in *Amoeba* by engulfment of material using cytoplasmic extensions called pseudopodia. • Ingestion of old red blood cells by Kupffer cells in the liver. • Ingestion of bacteria and cell debris by neutrophils and macrophages (phagocytic white blood cells).

3. Exocytosis examples (any of):
   • Secretion of substances from specialised secretory cells in multicellular organisms, e.g. hormones from endocrine cells, digestive secretions from exocrine cells. • Expulsion of wastes from unicellular organisms, e.g. *Paramecium* and *Amoeba* expelling residues from food vacuoles.

4. Any type of cytosis (unlike diffusion) is an active process involving the use of ATP. Low oxygen inhibits oxidative metabolism and lowers the energy yield from the respiration of substrates (ATP availability drops).

5. (a) **Oxygen**: Diffusion.
   (b) **Cellular debris**: Phagocytosis.
   (c) **Water**: Osmosis.
   (d) **Glucose**: Facilitated diffusion.

Active and Passive Transport (page 91)

1. **Passive transport** requires no energy input from the cell; materials follow a concentration gradient.
   Active transport requires considerable amounts of energy (ATP) to make materials go in a direction they would not normally go (at least at the rate required).
2. Gases moving by **diffusion**: Oxygen, carbon dioxide.
3. Cells in the digestive (exocrine) glands of the stomach, pancreas, upper small intestine (duodenum); endocrine glands (e.g. adrenal glands); salivary glands.
4. (a) Protozoan: *Amoeba*, *Paramecium*
   (b) A food vacuole develops at the end of the oral groove in *Paramecium* and is pinched off to circulate within the cell. In *Amoeba*, the pseudopodia engulf a food particle and a vacuole is formed where the membrane pinches off after the particle is engulfed.
   (c) Human cell: Phagocyte (phagocytic leucocyte).

Cell Division (page 93)

1. (a) Mitosis: Cell division for growth and repair produces cells with 2N chromosome number.
   (b) Meiosis: Cell division for producing gametes (sperm, pollen, eggs) with 1N chromosome number.
2. A **zygote** results from the fertilisation of an egg and sperm cell; it is diploid and gives rise (through mitosis and cellular differentiation) to a new individual.
3. In **spermatogenesis**, the nucleus of the germ cell divides twice to produce four similar sized gametes (sperm cells). In **oogenesis**, the two divisions are not equal and only one of the four nuclei (and most of the cytoplasm) produce the egg cell.

Mitosis and the Cell Cycle (page 94)

1. A. Anaphase
   B. Prophase
   C. Late metaphase (early anaphase is also acceptable).
   D. Late anaphase
   E. Cytokinesis (late telophase is also acceptable).
2. Replicate the DNA to form a second chromatid.
   Coil up into visible chromosomes to avoid tangling.
3. A. Interphase: The stage between cell divisions (mitoses). Just before mitosis, the DNA is replicated to form an extra copy of each chromosome (still part of the same chromosome as an extra chromatid).
   B. Late prophase: Chromosomes condense (coil and fold up) into visible form. Centrioles move to opposite ends of the cell.
   C. Metaphase: Spindle fibers form between the centrioles. Chromosomes attach to the spindle fibers at the cell 'equator'.
   D. Late anaphase: Chromatids from each chromosome are pulled apart and move in opposite directions, towards the centrioles.
   E. Telophase: Chromosomes begin to unwind again. Two new nuclei form. The cell plate forms across the midline where the new cell wall will form.
   F. Cytokinesis: Cell cytoplasm divides to create two distinct 'daughter cells' from the original cell. It is in this form for most of its existence, and carries out its designated role (normal function).

The Genetic Origins of Cancer (page 96)

1. Cancerous cells have lost control of the genetic mechanisms regulating the cell cycle so that the cells become immortal. They also lose their specialised functions and are unable to perform their roles.
2. The cell cycle is normally controlled by two types of gene: proto-oncogenes, which start cell division and are required for normal cell development, and **tumour-suppressor genes**, which switch cell division off. Tumour suppressor genes will also halt cell division if the DNA is damaged and, if the damage is not repairable, will bring about a programmed cell suicide (apoptosis).
3. Normal controls over the cell cycle can be lost if either the proto-oncogenes or the tumour suppressor genes acquire mutations. Mutations to the proto-oncogenes, with the consequent formation of **oncogenes**, results in uncontrolled cell division. Mutations to the tumour-suppressor genes results in a failure to regulate the cell repair processes and a failure of the cell to stop dividing when damaged.

Cell Growth and Cancer (page 97)

1. Exposure to carcinogens can damage the DNA and trigger uncontrolled cell division (and tumour formation).
2. A single cause of cancer can be difficult to pin-point because there are many factors (environmental, lifestyle, genetic, and ageing) which can interact and result in the development of a cancer.

Stem Cells (page 98)

1. (a) Self renewal while remaining unspecialised.
   (b) Potency, the ability to differentiate into any number of specialised cells
2. Embryonic stem cells are pluripotent and can form any cells derived from the three germ layers. Adult stem cells are multipotent; they are more limited in the numbers of cells that they can form, and generally form the cells of the blood, heart, muscle, and nerves. Embryonic stem cells have far greater medical potential because of their ability to form all cell types. They potentially could be used for regenerative medicine and tissue replacement.
3. Adult stem cells are often involved in replacement of dying cells and the repair of damaged tissue. Students may provide any one of the following examples to illustrate the role of adult stem cells:
   - Hematopoietic stem cells,from bone marrow, regenerate all blood cell types to replace old or damaged cells.
   - Bone marrow stromal cells (mesenchymal stem cells) give rise to a variety of cell types: bone cells (osteocytes), cartilage cells (chondrocytes), fat cells (adipocytes), and other kinds of connective tissue cells such as those in tendons.
   - Neural stem cells in the brain give rise to three major cell types: neurones, astrocytes, and oligodendrocytes.
   - Epithelial stem cells in the lining of the digestive tract are continually producing cells to replace those damaged during the digestive process. Cell

types include absorptive cells, goblet cells, Paneth cells, and enteroendocrine cells.
- Skin stem cells give rise to keratinocytes, which migrate to the surface of the skin and are regenerated frequently to replace old or damaged cells and maintain a protective layer.

Differentiation of Human Cells (page 99)

1. 230 different cell types
2. 50 cell divisions
3. 100 billion cells
4. Skin cells, intestinal lining cells, blood (stem) cells
5. Nerve cells, bone cells, kidney cells
6. (a) Germ line is the series of cell divisions destined to produce gamete cells.
   (b) Germ cells will produce gametes (eggs and sperm) and must be essentially unspecialised cells. This is necessary so that none of the genes that are needed to produce the 230 specialised cells in new offspring are turned off before they are needed.
7. (a) A clone is a copy of a cell (or complete organism) with a genetic makeup that is identical to the single parent cell it was created from.
   (b) As for 6(b): None of the genes required to produce specialised cells have been turned off.
8. Cancerous cells are those that have lost control of the regulatory processes that govern the cell's function. Instead they become generalised cells that lose their tissue identity, pulling away from cells around them and undergoing cell division at a rapid rate.
9. At certain stages in the sequence of cell divisions as the embryo grows, some genes get switched on while others get switched off permanently, causing the cells to take on specialised functions.

Human Cell Specialisation (page 101)

1. (b) **Erythrocyte**:
   Features: Biconcave cell, lacking mitochondria, nucleus, and most internal membranes. Contains the oxygen-transporting pigment, haemoglobin.
   Role: Uptake, transport, and release of oxygen to the tissues. Some transport of CO_2. Lack of organelles creates more space for oxygen transport. Lack of mitochondria prevents oxygen use.
   (c) **Retinal cell**:
   Features: Long, narrow cell with light-sensitive pigment (rhodopsin) embedded in the membranes.
   Role: Detection of light: light causes a structural change in the membranes and leads to a nerve impulse (result is visual perception).
   (d) **Skeletal muscle cell(s)**:
   Features: Cylindrical shape with banded myofibrils. Capable of contraction (shortening).
   Role: Move voluntary muscles acting on skeleton.
   (e) **Intestinal epithelial cell(s)**:
   Features: Columnar cell with a high surface area as a result of fingerlike projections (microvilli).
   Role: Absorption of digested food.
   (f) **Motor neurone cell**:
   Features: Cell body with a long extension (the axon) ending in synaptic bodies. Axon is insulated with a sheath of fatty material (myelin).
   Role: Rapid conduction of motor nerve impulses from the spinal cord to effectors (e.g. muscle).
   (g) **Spermatocyte**:
   Features: Motile, flagellated cell with mitochondria. Nucleus forms a large proportion of the cell.
   Role: Male gamete for sexual reproduction. Mitochondria provide the energy for motility.
   (h) **Osteocyte**:
   Features: Cell with calcium matrix around it. Fingerlike extensions enable the cell to be supplied with nutrients and wastes to be removed.
   Role: In early stages, secretes the matrix that will be the structural component of bone. Provides strength.

Plant Cell Specialisation (page 102)

1. (b) **Pollen grain**:
   Features: Small, lightweight, often with spikes.
   Role: houses male gamete for sexual reproduction.
   (c) **Palisade parenchyma cell**:
   Features: Column-shaped cell with chloroplasts.
   Role: Primary photosynthetic cells of the leaf.
   (d) **Epidermal cell**:
   Features: Waxy surface on a flat-shaped cell.
   Role: Provides a barrier to water loss on leaf.
   (e) **Vessel element**:
   Features: Rigid remains of a dead cell. No cytoplasm. End walls perforated. Walls are strengthened with lignin fibres.
   Role: Rapid conduction of water through the stem. Provides support for stem/trunk.
   (f) **Stone cell**:
   Features: Very thick lignified cell wall inside the primary cell wall. The cytoplasm is restricted to a small central region of the cell.
   Role: Protection of the seed inside the fruit.
   (g) **Sieve tube member**:
   Features: Long, tube-shaped cell without a nucleus. Cytoplasm continuous with other sieve cells above and below it. Cytoplasmic streaming is evident.
   Role: Responsible for translocation of sugars etc.
   (h) **Root hair cell:**
   Features: Thin cuticle with no waxy layer. High surface area relative to volume.
   Role: Facilitates the uptake of water and ions.

Root Cell Development (page 103)

1. (a) Cells specialise to take on specific functions.
   (b) Cells are becoming longer and/or larger.
   (c) Cells are dividing by mitosis.
2. (a) Late anaphase; chromatids are being pulled apart and are at opposite poles.
   (b) Telophase; there are two new nuclei formed and the cell plate is visible.
   (c) 25 of 250 cells were in mitosis, therefore mitosis occupies 25/250 or one tenth of the cell cycle.
3. The **cambium layer** of cells (lying under the bark between the outer phloem layer of cells and the inner xylem layer of cells). **Note**: Cells dividing from each side of this layer specialise to form new phloem on the outside and new xylem on the inside.

Levels of Organisation (page 104)

1. **Animals**
   (a) **Organ system**: Nervous system, reproductive system
   (b) **Organs**: Brain, heart, spleen
   (c) **Tissues**: Blood, bone, cardiac muscle, cartilage, squamous epithelium
   (d) **Cells**: Leukocyte, mast cell, neuron, Schwann cell
   (e) **Organelles**: Lysosome, ribosomes
   (f) **Molecular**: Adrenaline, collagen, DNA, phospholipid

2. **Plants**
   (a) **Organs**: Flowers, leaf, roots
   (b) **Tissues**: Collenchyma*, mesophyll, parenchyma*, phloem, sclerenchyma
   (c) **Cells**: Companion cells, epidermal cell, fibers, tracheid
   (d) **Organelles**: Chloroplasts, ribosomes
   (e) **Molecular**: Pectin, cellulose, DNA, phospholipid

   * **Note**: Parenchyma and collenchyma are simple tissues comprising only one type of cell (parenchyma and collenchyma cells respectively). Simple plant tissues are usually identified by cell name alone.

Animal Tissues (page 105)

1. The organisation of cells into specialized tissues allows the tissues to perform particular functions. This improves efficiency of function because different tasks can be shared amongst specialized cells. Energy is saved in not maintaining non-essential organelles in cells that do not require them.

2. (a) **Epithelial tissues**: Single or multiple layers of simple cells forming the lining of internal and external body surfaces. Cells rest on a basement membrane of fibers and collagen and may be specialized. **Note**: epithelial cells may be variously shaped: squamous (flat), cuboidal, columnar etc.
   (b) **Nervous tissue**: Tissue comprising densely packed nerve cells specialized for transmitting electro-chemical impulses. Nerve cells may be associated with supportive cells (e.g. Schwann cells), connective tissue, and blood vessels.
   (c) **Muscle tissue**: Dense tissue comprising highly specialized contractile cells called fibers held together by connective tissues.
   (d) **Connective tissues**: Supporting tissue of the body, comprising cells widely dispersed in a semi-fluid matrix (or fluid in the case of blood and lymph).

3. (a) Muscle tissue is made up of long muscle fiber cells and myobirils which are made up of contractile proteins actin and myosin. These allow the muscle fibers to contact when stimulated. The contraction results in movement of the organism itself (locomotion) or movement of an internal organ.
   (b) Nervous tissue is comprised of two main tissue types, neurons which transmit nerve signals and glial cells which provide support to the neurons. Neurons have several protrusions (dendrites or axons) from their cell body which allow conduction of nerves impulses to target cells.

Plant Tissues (page 106)

1. **Collenchyma**
   Cell type(s): collenchyma cells
   Role: provides flexible support.

   Sclerenchyma
   Cell type(s): sclerenchyma cells
   Role: provides rigid, hard support.

   Root Endodermis
   Cell type(s): endodermal cells
   Role: Provides selective barrier regulating the passage of substances from the soil to the vascular tissue.

   Pericycle
   Cell type(s): parenchyma cells
   Role: Production of branch roots, synthesis and transport of alkaloids.

   Leaf mesophyll
   Cell type(s): spongy mesophyll, palisade mesophyll
   Role: Main photosynthesis site in the plant.

   Xylem
   Cell type(s): tracheids, vessel members, fibers, paraenchyma cells
   Role: Conducts water and dissolved minerals in vascular plants.

   Phloem
   Cell type(s): sieve-tube members, companion cells, parenchyma, fibers, sclereids
   Role: transport of dissolved organic material (including sugars) within vascular plants.

   Epidermis
   Cell type(s): epidermal cells, guard cells, subsidiary cells, and epidermal hairs (trichomes).
   Role: Protection against water loss, regulation of gas exchange, secretion, water and mineral absorption.

Introduction to Gas Exchange (page 108)

1. **Cellular respiration** refers to the production of ATP through the oxidation of glucose. **Gas exchange** refers to the way in which respiratory gases (oxygen and carbon dioxide) are exchanged with the environment. Oxygen is required to drive the reactions of cellular respiration. Carbon dioxide is a waste product.

2. (a) Moist so that gases can dissolve and diffuse across.
   (b) Large surface area to provide for a large amount of gas exchange (to meet the organism's needs).
   (c) Thin membrane that does not present a large barrier to diffusion of gases. This provides a surface across which gases easily diffuse.

3. The rate of diffusion across the gas exchange surface will be more rapid when membrane surface area or concentration difference across the membrane increases and/or membrane thickness decreases.

Gas Exchange in Animals (page 109)

1. (a) Provides adequate supply and removal of respiratory gases necessary for an active (metabolically demanding) lifestyle.
   (b) Enables animals to attain a larger size (as they are freed from a dependence on direct diffusion of gases across thin body surfaces).

2. (a) The air sacs function in ventilating the lungs (where gas exchange takes place). They facilitate one way

(rather than to and fro) flow of air through the lungs.
(b) Birds require an efficient gas exchange system because of their high metabolic rate (associated with flight). However they do not want to carry a large amount of lung tissue because this would be heavy and hinder flight (hence air sacs).

3. (a) **Body surface**
Location: The entire body surface is involved.
Group: Characteristic of small and/or thin animals, e.g. cnidarians, ctenophores, annelids, flatworms.
Medium: Air (in damp environments) or water.
(b) **Tracheal tubes**
Location: Thin tubes extend inwards from spiracles at the body surface located on the abdomen.
Group: Insects and some spiders.
Medium: Air.
(c) **Gills**
Location: Thin, filamentous structures that extend outside the main body from behind the head/buccal area in vertebrates or associated with the thorax, abdomen or limbs in invertebrates.
Group: Fish and most crustaceans.
Medium: Water.
(d) **Lungs**
Location: Invaginations (in-pockets) of the body surface (inside body) within the thoracic region.
Group: Vertebrates other than fish.
Medium: Air.

4. (a) Air breathers produce mucus that keeps the gas exchange surface moist.
(b) Some water vapour is present in lungs as a result of metabolism.

5. Large amounts of organic material clog the gill surface and prevent the water closely contacting it. Organic material also consumes a lot of oxygen in its decomposition. This reduces the amount of oxygen in the water available to animals for gas exchange.

6. An animal's gas exchange system must be appropriate to the environment in which it must operate. Gills do not function in air because the gill tissue needs to be supported by the water to prevent its collapse in the less dense medium of air. In air, the gill tissue rapidly dries out and, once dry, the surface will not operate effectively for gas exchange. In water, lungs do not function because water is too dense a medium to enter and leave an internalised structure. The tracheae of insects operate well in terrestrial organisms of a small size because they can provide oxygen directly to the tissues. With direct oxygen delivery, a respiratory pigment in the blood is not required. In aquatic insects, the tracheae extend into flattened gills on the abdomen, and these increase oxygen uptake from the water, where oxygen extraction is more difficult than in air (because diffusion rates are slower).

The Human Respiratory System (page 111)

1. (a) The structural arrangement (lobes, each with its own bronchus and dividing many times before terminating in numerous alveoli) provides an immense surface area for gas exchange.
(b) Gas exchange takes place in the alveoli.

2. The respiratory membrane is the layered junction between the alveolar cells, the endothelial cells of the capillaries, and their associated basement membranes. It provides a surface across which gases can freely move.

3. Surfactant reduces the surface tension of the lung tissue and counteracts the tendency of the alveoli to recoil inward and stick together after each expiration.

4. Completed table as below:

	Region	Cartilage	Ciliated epithelium	Goblet cells (mucus)	Smooth muscle	Connective tissue
1	Trachea	✓	✓	✓	✓	✓
2	Bronchus	✓	✓	✓	✓	✓
3	Bronchioles	gradually lost	✓	✓	✓	✓
4	Alveolar duct	✗	✗	✗	✓	✓
5	Alveoli	✗	✗	✗	very little	✓

5. Respiratory distress syndrome: The lack of surfactant and high surface tension in the alveoli result in the collapse of the lungs to an uninflated state after each breath. Breathing is difficult and laboured, oxygen delivery is inadequate and, if untreated, death usually follows in a few hours.

Breathing in Humans (page 113)

1. (a) **Quiet breathing**: External intercostal muscles and diaphragm contract. Lung space increases and air flows into the lungs (inspiration). Inflation is detected and breath in ends. Expiration occurs through elastic recoil of the ribcage and lung tissue (air flows passively out to equalise with outside air pressure).
(b) During forced or **active breathing**, muscular contraction is involved in both the inspiration and the expiration (expiration is not passive).

2. (a) Tidal volume vol: 0.5 dm^3
(b) Expiratory reserve volume vol: 1.0 dm^3
(c) Residual volume vol: 1.2 dm^3
(d) Inspiratory capacity vol: 3.8 dm^3
(e) Vital capacity vol: 4.8 dm^3
(f) Total lung capacity vol: 6.0 dm^3

3. **G**: Tidal volume is increasing as a result of exercise.

4. PV: 15 x 400 = 6000 cm^3 or 6 dm^3

5. (a) There is 90X more CO_2 in exhaled air than in inhaled air (3.6 ÷ 0.04).
(b) The CO_2 is the product of cellular respiration in the tissues. **Note**: Some texts give a value of 4.0% for exhaled air (100X the CO_2 content of inhaled air).
(c) The dead space air is not involved in gas exchange therefore retains a higher oxygen content than the air that leaves the alveoli air. This raises the oxygen content of the expired air.

Control of Breathing (page 115)

1. The basic rhythm of breathing is controlled by the respiratory centre in the medulla which sends rhythmic impulses to the intercostal muscles and diaphragm to bring about normal breathing.

2. (a) **Phrenic nerve**: Innervates the diaphragm (which

contracts and moves down in inspiration).
- (b) **Intercostal nerves**: Innervate the intercostal muscles (internal and external intercostal nerves and muscles) to bring about movements of the ribcage.
- (c) **Vagus nerve**: Sensory portion carries impulses from stretch receptors in the bronchioles to the respiratory centre to inhibit inspiration (called the inflation reflex).
- (d) **Inflation reflex** (also known as the **Hering-Breuer reflex)**: The inhibition of the inspiratory centre to end the breath in. **Note**: Sensory impulses from the stretch receptors in the bronchioles travel (via the vagus) to inhibit the inspiratory centre and expiration follows. When the lungs deflate, the stretch receptors are not stimulated and the inhibition of the inspiratory centre stops.

3. (a) Low blood pH increases rate and depth of breathing.
   (b) Sensory information from aortic and carotid chemoreceptors (bodies) is sent to the respiratory centre, which mediates the increase in breathing rate. **Note**: Sensory impulses are sent from the carotid bodies (chemoreceptors) via the carotid sinus nerve and then the glossopharyngeal nerve (IX cranial). Sensory impulses from the aortic bodies (chemoreceptors) travel in the vagus nerve. Low blood pH also stimulates the chemosensitive area in the medulla directly.
   (c) Blood pH is a good indicator of high carbon dioxide levels (and therefore a need to increase respiratory rate to remove the CO_2 and obtain more oxygen).

Review of Lung Function (page 116)

1. (a) Nasal cavity
   (b) Oral cavity
   (c) Trachea
   (d) Lung
   (e) Terminal bronchiole
   (f) Alveoli
   (g) Diaphragm
   (i) Medullary respiratory centre
   (ii) Vagus nerve
   (iii) Intercostal nerve
   (iv) Chemoreceptors
   (v) Stretch receptors
   (vi) Phrenic nerve

2. A = Inspiratory reserve volume vol: 3.3 dm^3
   B = Inspiratory capacity vol: 3.8 dm^3
   C = Tidal volume vol: 0.5 dm^3
   D = Expiratory reserve volume vol: 1.0 dm^3
   E = Residual volume vol: 1.2 dm^3

Respiratory Pigments (page 117)

1. (a) Respiratory pigments are able to bind reversibly with oxygen. They may bind and carry several oxygen molecules (and therefore increase the amount that can be carried over what can be dissolved in the plasma, which is very low).
   (b) The number of metal-containing prosthetic groups.

2. Organisms with a high metabolic activity (therefore high oxygen demand) have haemoglobins with a greater oxygen carrying capacity (values are highest in endothermic homeotherms, i.e. birds and mammals).

3. Large molecular weight respiratory pigments are too large to be held within cells and must be carried dissolved in the plasma.

Internal Transport (page 119)

1. Diffusion is too inefficient and slow to provide raw materials quickly enough to all the cells of larger animals. The materials must be transported to the places where they are needed.

2. (a) Blood vessels carry the blood, transporting it to (or almost to) where it is required.
   (b) The heart pumps the blood (or circulatory fluid) to different parts of the body.
   (c) The blood or haemolymph is a medium to carry the nutrients, wastes, and usually gases to and from places in the body.

3. Simple terrestrial organisms can use diffusion if they maintain a moist body surface (e.g. by mucus secretion).

Mammalian Transport (page 120)

1. (a) Head
   (b) Lungs
   (c) Liver
   (d) Gut (intestines)
   (e) Kidneys
   (f) Genitals/lower body

Circulatory Systems (page 121)

1. In a **closed** system the blood is enclosed entirely within vessels. In an **open** system, a lot of the blood is free within large spaces (sinuses) where it bathes the cells.

2. A double circulation system is more efficient because, after being oxygenated, the blood returns to the heart and is then pumped to the body tissues. This allows a higher blood pressure to be maintained and a faster delivery of oxygen and nutrients to the body tissues.

3. (a) **Fish**: Three chambers in series. The **sinus venosus** receives blood returning to the heart from the body's tissues, blood is then pumped into the single **atrium**, and then into the single **ventricle** (the main pumping force). The ventricle pumps the blood on to the gills. **Note**: A fourth region in the fish heart, the **conus arteriosus**, is located at the ventricular end of the heart (in series). Functionally it is a fourth chamber, but is often not referred to at a secondary school level of anatomy. Note that the fish heart is part of a single circuit system, with no separate pulmonary circuit.
   (b) **Amphibians**: A double circuit system (systemic and pulmonary circuits) with three distinct chambers. The **left atrium**, receiving blood from the pulmonary circuit, lies beside the **right atrium**, which receives blood from the systemic (body) circuit. Both atria pump blood into the single ventricle, which pumps the (mixed) blood to the body and lungs. **Note**: In amphibians, the sinus venosus (large in fish) is still a prominent feature, although functionally it becomes an extension of the right atrium. It is found on the dorsal surface of the heart and is thus not evident on the diagram.
   (c) **Mammals**: A double circuit system with four chambers. The **left atrium**, receiving blood from the pulmonary circuit, pumps blood into the **left ventricle**, which pumps the oxygenated blood through the systemic (body) circuit. The **right atrium** (beside, but separate from, the left atrium) receives deoxygenated blood from the body and pumps it into the **right ventricle** (beside, but separate from, the left ventricle), which pumps blood

to the lungs (through the pulmonary circuit).

4. After passing through the gills, the blood moves through the systemic circulation (the body's tissues).

Arteries (page 123)

1. (a) Tunica externa
   (b) Tunica media
   (c) Endothelium
   (d) Blood (or lumen)

2. (a) Thick, elastic walls can withstand the high pressure of the blood being pumped from the heart. **Note**: Elasticity also helps to even out the surges that occur with each contraction of the heart. This keeps the blood moving forward in a continuous flow.
   (b) Blood pressure is low within the arterioles.

3. The smooth muscle around arteries helps to regulate blood flow and pressure. By contracting or relaxing it alters the diameter of the artery and adjusts the volume of blood as required.

4. (a) The diameter of the artery increases.
   (b) The blood pressure decreases.

Veins (page 124)

1. (a) Veins have less elastic and muscle tissue than arteries.
   (b) Veins have a larger lumen than arteries.

2. Most of the structural differences between arteries and veins are related to the different blood pressures inside the vessels. Blood in veins travels at low pressure and veins do not need to be as strong, hence the thinner layers of muscle and elastic tissue and the relatively larger lumen. **Note**: There is still enough elastic and muscle tissue to enable the veins to adjust to changes in blood volume and pressure.

3. Veins are "massaged" by the skeletal muscles (e.g. leg muscles). Valves (together with these muscular movements) help to return venous blood to the heart by preventing backflow away from the heart. **Note**: When skeletal muscles contract and tighten around a vein the valves open and blood is driven towards the heart. When the muscles relax, the valves close, preventing backflow.

4. Venous blood oozes out in an even flow from a wound because it has lost a lot of pressure after passing through the narrow capillary vessels (with their high resistance to flow). Arterial blood spurts out rapidly because it is being pumped directly from the heart and has not yet entered the capillary networks.

Capillaries and Tissue Fluid (page 125)

1. **Capillaries** are very small blood vessels forming networks or beds that penetrate all parts of the body. The only tissue present is an endothelium of squamous epithelial cells. In contrast, **arteries** have a thin endothelium, a central layer of elastic tissue and smooth muscle and a thick outer layer of elastic and connective tissue. **Veins** have a thin endothelium, a central layer of elastic and muscle tissue and a thin outer layer of elastic connective tissue. In addition, veins also have valves.

2. (a) Sinusoids differ from capillaries in that they are wider and follow a more convoluted path through the tissue. They are lined with phagocytic cells rather than the usual endothelial lining of capillaries.
   (b) Capillaries and sinusoids are similar in that they both transport blood from arterioles to venules.

3. (a) Leakage of fluid from capillaries produces tissue fluid, which bathes the tissues, providing oxygen and nutrients as well as a medium for the transport (away) of metabolic wastes, e.g. CO_2.
   (b) Capillary walls are thin enough to allow exchanges. No exchange occurs in arteries and veins because the walls of arteries and veins are too thick.

4. (a) Arteriolar end: Hydrostatic pressure predominates in causing fluid to move out of the capillaries.
   (b) Venous end: Increased concentration of solutes and reduction in hydrostatic pressure at the venous end of a capillary bed **lowers the solute potential** within the capillary and there is a tendency for water and solutes to re-enter the capillary.

5. (a) Most tissue fluid finds it way directly back into the capillaries as a result of net inward pressure at the venule end of the capillary bed.
   (b) The lymph vessels (which parallel the blood system) drain tissue fluid (as lymph) back into the heart, thereby returning it into the main circulation.

The Effects of High Altitude (page 127)

1. (a) Less oxygen is available for metabolic activity so people become breathless and often dizzy. Associated effects are headache, nausea, tiredness and coughing.
   (b) Altitude sickness or mountain sickness.

2. (a) Heart and breathing rates increase.
   (b) Increased breathing rate increases the rate at which new air is brought into the lungs (compensating for lower oxygen). Increased heart rate pumps blood more rapidly to tissues to improve oxygen delivery.

3. (a) Any of:
   - The concentration of red blood cells in the blood may increase.
   - The blood may become thicker.
   - Over a long period of time, capillary networks increase in density.

   (b) Any of:
   - Increased RBCs and blood concentration increase the amount of oxygen that can be carried in the body.
   - Increased capillary density increases the amount and efficiency of oxygen delivery to the tissues.

Exercise and Blood Flow (page 128)

1. Answers for missing values are listed from top to bottom under the appropriate heading:

	At rest (% of total)	**Exercise** (% of total)
Heart	4.0	4.2
Lung	2.0	1.1
Kidneys	22.0	3.4
Liver	27.0	3.4
Muscle	15.0	70.2
Bone	5.0	1.4

Skin	6.0	10.7
Thyroid	1.0	0.3
Adrenals	0.5	0.1
Other	3.5	1.0

2. The heart beats faster and harder to increase the volume of blood pumped per beat and the number of beats per minute (increased blood flow).

3. (a) Blood flow increases approximately 3.5 times.
   (b) Working tissues require more oxygen and nutrients than can be delivered by a resting rate of blood flow. Therefore the rate of blood flow (delivery to the tissues) must increase during exercise.

4. (a) Thyroid and adrenal glands, as well as the tissues other than those defined in the table, show no change in absolute rate of blood flow.
   (b) This is because they are not involved in exercise and do not require an increased blood flow. However, they do need to maintain their usual blood supply and cannot tolerate an absolute decline.

5. (a) Skeletal muscles (increases 16.7X), skin (increases 6.3X), and heart (increases 3.7X)
   (b) These tissues and organs are all directly involved in the exercise process and need a greater rate of supply of oxygen and nutrients. Skeletal muscles move the body, the heart must pump a greater volume of blood at a greater rate and the skin must help cool the body to maintain core temperature.

Blood (page 129)

1. **Note**: In some cases, the answers below provide more detail than expected. This is provided as extension.
   (b) Protection against disease:
   Blood component: White blood cells
   Mode of action: Engulf bacteria, mediate immune reactions, and allergic and inflammatory responses.
   (c) Communication between cells, tissues and organs:
   Blood component: Hormones
   Mode of action: Specific chemicals which are carried in the blood to target tissues, where they interact with specific receptors and bring about an appropriate response.
   (d) Oxygen transport:
   Blood component: Haemoglobin molecule of erythrocytes.
   Mode of action: Binds oxygen at the lungs and releases it at the tissues.
   (e) Carbon dioxide transport:
   Blood components: Mainly plasma (most carbon dioxide is carried as bicarbonate in the plasma, a small amount is dissolved in the plasma). Red blood cells (a small amount (10-20%) of carbon dioxide is carried bound to haemoglobin).
   Mode of action: Diffuses between tissues, plasma, and lungs according to concentration gradient.
   (f) Buffer against pH changes:
   Blood components: Haemoglobin molecule of erythrocytes. Plasma bicarbonate and proteins.
   Mode of action: Free hydrogen ions are picked up and carried by the haemoglobin molecule (removed from solution). Plasma bicarbonate can form either carbonic acid by picking up a hydrogen ion (H^+), or sodium bicarbonate by combining with sodium ions. Negatively charged proteins also associate with H^+.
   (g) Nutrient supply:
   Blood component: Plasma
   Mode of action: Glucose is carried in the plasma and is taken up by cells (made available throughout the body to all tissues).
   (h) Tissue repair:
   Blood components: Platelets and leucocytes
   Mode of action: Platelets initiate the cascade of reactions involved in clotting and wound repair. Leucocytes (some types) engulf bacteria and foreign material, preventing or halting infection.
   (i) Transport of hormones, lipids, and fat soluble vitamins:
   Blood component: α-globulins
   Mode of action: α-globulins bind these substances and carry them in the plasma. This prevents them being filtered in the kidneys and lost in the urine.

2. Any of: Presence (WBC) or absence (RBC) of **nucleus**. Colour, reflecting presence (RBC) or absence (WBC) of respiratory pigment, **haemoglobin**. **Shape and size** (smaller, dish shaped RBCs vs larger, rounded WBCs. **Mitochondria** present in WBCs, absent in RBCs.

3. (a) Lack of a nucleus allows more space inside the cell to carry Hb (hence greater O_2 carrying capacity).
   (b) Lack of mitochondria forces the red blood cells to metabolise anaerobically so that they do not consume the oxygen they are carrying.

4. (a) Elevated eosinophil count: Allergic response such as hay fever or asthma.
   (b) Elevated neutrophil count: Microbial infection.
   (c) Elevated basophil count: Inflammatory response e.g. as a result of an allergy or a parasitic (as opposed to bacterial) infection.
   (d) Elevated lymphocyte count: Infection or response to vaccination.

Gas Transport in Humans (page 131)

1. (a) Oxygen is high in the lung alveoli and in the capillaries leaving the lung.
   (b) Carbon dioxide is high in the capillaries leaving the tissues and in the cells of the body tissues.

2. Haemoglobin binds oxygen reversibly, taking up oxygen when oxygen tensions are high (lungs), carries oxygen to where it is required (the tissues) and releases it.

3. (a) As oxygen level in the blood increases, more oxygen combines with haemoglobin. However, the relationship is not linear: Hb saturation remains high even when blood oxygen levels fall very low.
   (b) When oxygen level (partial pressure) in the blood or tissues is low, haemoglobin saturation declines markedly and oxygen is released (to the tissues).

4. (a) Foetal Hb has a higher affinity for oxygen than adult Hb (it can carry 20-30% more oxygen).
   (b) This higher affinity is necessary because it enables oxygen to pass from the maternal Hb to the foetal Hb across the placenta.

5. (a) The Bohr effect
   (b) Actively respiring tissue (especially tissue with high metabolic demand, such as working muscle) consumes a lot of oxygen and generates a lot of carbon dioxide. This lowers tissue (blood) pH causing more oxygen to be released from the

haemoglobin to where it is required.

6. Myoglobin preferentially picks up oxygen from Hb and is able to act as an oxygen store in the muscle.

7. Any two of: **Haemoglobin**, which picks up H^+ generated by the dissociation of carbonic acid. **Bicarbonate** alone (from this dissociation), and combined with Na^+ (from the dissociation of NaCl). **Blood proteins**.

The Human Heart (page 133)

1. (a) Pulmonary artery (e) Aorta
   (b) Vena cava (f) Pulmonary vein
   (c) Right atrium (g) Left atrium
   (d) Right ventricle (h) Left ventricle

Positions of heart valves

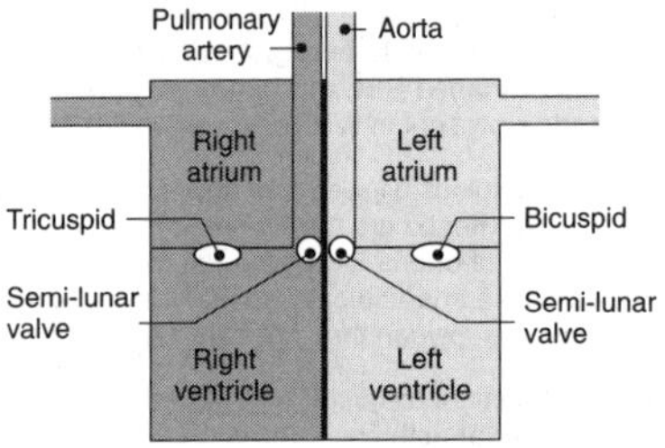

2. **Valves** prevent the blood from flowing the wrong way through the heart and help regulate filling of the chambers.

3. (a) The heart has its own coronary blood supply to meet the high oxygen demands of the heart tissue.
   (b) There must be a system within the heart muscle itself to return deoxygenated blood and waste products of metabolism back to the right atrium.

4. If blood flow to a particular part of the heart is restricted or blocked (because of blocked blood vessel), the part of the heart muscle supplied by that vessel will die, leading to a heart attack or infarction.

5. A: arterioles B: venules
   C: arterioles D: capillaries

6. (a) The **pulse pressure** is the difference between the systolic pressure and the diastolic pressure.
   (b) Pulse pressure between the aorta and the capillaries will decrease (because of the increasing resistance met on route).

7. (a) You are recording expansion and recoil of the artery that occurs with each contraction of the left ventricle.
   (b) The best place to take a **pulse** is from the brachial or carotid artery. The blood flow from these arteries (close to the heart) is at a high pressure and still carries the beat or pulse.

Control of Heart Activity (page 135)

1. (a) **Sinoatrial node**: Initiates the cardiac cycle through the spontaneous generation of action potentials.
   (b) **Atrioventricular node**: Delays the impulse.
   (c) **Bundle of His**: Distributes the action potentials over the ventricles (resulting in ventricular contraction).

2. Delaying the impulse at the AVN allows time for atrial contraction to finish before the ventricles contract.

3. (a) Cardiac output
   $70\ cm^3$ X 70 beats per min. = $4900\ cm^3$ = $4.9\ dm^3$.
   (b) Trained endurance athletes achieve a high cardiac output primarily through having a very high stroke volume: the heart pumps a lot of blood with each beat. For any given level of exercise, their heart rates are relatively low.

4. (a) **Myogenic**: The heart muscle is capable of rhythmic contraction independently of any external nervous stimulation.
   (b) **Evidence**: When the heart is removed from its nervous supply (if provided with adequate oxygen, ions, and fluids) will continue to beat.

5. Heavy exercise alters blood composition (lowering blood oxygen and pH) and increases venous return to the heart. These changes stimulate the accelerator centre in the medulla either directly (via blood composition) or via sensory impulses from stretch receptors in the vena cavae and atrium (venous return). The accelerator centre responds by sympathetic stimulation of the heart (via the cardiac nerve and release of noradrenaline) bringing about increased rate and force of heart contraction.

6. (a) Increased arterial flow (in aorta and carotid arteries).
   (b) Stretch receptors in carotid sinus and aorta detect an increase in arterial flow and send afferent impulses to the inhibitory centre in the medulla. The inhibitory centre mediates a decrease in heart rate via the vagus nerve (the carotid and aortic reflexes). There is a subsequent decrease in cardiac output and arterial blood pressure.

7. (a) and (b) any of the following in any order:
   - Aortic pressure receptors (baroreceptors) in the wall of the aortic arch respond to increased arterial blood flow in the aorta.
   - Carotid baroreceptors in the carotid sinus respond to increased arterial flow in the carotid artery.
   - Baroreceptors in the vena cava and the right atrium respond to increased venous return (mediates an increase in heart rate - the Bainbridge reflex).

8. (a) Individuals have different tolerances to the same level of a substance (also depends on their previous intakes of other substances such as coffee). Different body mass will affect the response also.
   (b) Controls should drink an energy drink that does not contain guarana.

The Cardiac Cycle (page 137)

1. (a) QRS complex (b) T (c) P

2. During the period of electrical recovery the heart muscle cannot contract. This ensures that the heart has an enforced rest and will not fatigue, nor accumulate lactic acid (as occurs in working skeletal muscle).

Review of the Human Heart (page 138)

1. (A) Cardiac nerve (sympathetic nerve is acceptable).
   (B) Baroreceptors in the vena cava and right atrium.
   (C) Vagus nerve (parasympathetic nerve is acceptable).

(D) Aortic baroreceptors.
(1) Sinoatrial node (SAN).
(2) Atrioventricular node (AVN).
(3) Bundle of His (atrioventricular bundle).
(4) Right ventricle.
(5) Pulmonary artery.
(6) Left atrium.
(7) Purkinje fibres.
(8) Left ventricle.

2. A = Contraction of the atria.
   B = Contraction of the ventricles.
   C = Relaxation and recovery of ventricles.

3. Use of an artificial pacemaker to regulate heart rhythm.

Transport in Plants (page 140)

1. (a) **Leaves**: Collect the sun's energy and convert it to usable energy (sugar). Control the entry and exit of gases and water vapour (therefore have a role in cooling the plant).
   Roots: Anchor the plant. Absorb water and dissolved minerals. Sometimes store food. Produce new tissue at meristems.
   Stems: Link the roots to the leaves. Provide support for the leaves, fruits, and flowers. Conduct water and dissolved minerals and foods around the plant. Produce new tissue at meristems.
   (b) Materials transported: Water, minerals (e.g. nitrogen, phosphorus), sugar, essential ions (e.g. K^+, Na^+).
   (c) Functions of specific transport tissues:
   Xylem: Transports water and dissolved minerals, plant support. **Phloem**: Transports primarily sugar (in solution), but also minerals, hormones, and amino acids, plant support.

2. Water

3. (a) Xylem: Osmosis (passive movement of water across a partially permeable membrane from less negative to more negative water potential / lower to higher solute concentration). Also cohesion and root pressure.
   (b) Phloem: Active transport to load the sugar into the phloem tissue, osmosis (water follows sugar into the phloem), pressure-flow of the sap along the phloem.

Stems and Roots (page 141)

1. A: Epidermis
   B: Fibre cap
   C: Phloem
   D: Cortex
   E: Vascular cambium
   F: Xylem
   G: Pith

2. Stems are distinguished by the presence of nodes and internodes.

3. Either of the following:
   – Dicot stems have a large central pith composed of parenchyma cells.
   – The vascular bundles are arranged in an orderly fashion around the periphery of the stem.

4. The cells of the vascular cambium divide to produce the thickening of the stem.

5. Any three of the following:
   – Roots anchor the plant into the soil
   – Absorb water and inorganic nutrients from the soil.
   – Sites for production of hormones gibberellins and cytokinins, which influence growth and development.
   – Specialised roots can have a variety of roles including supporting stems (prop roots), or supplying oxygen to underwater roots (pneumatophores) in the case of mangroves.

6. (a) and (b), any two of:
   - The primary xylem forms a star shape in the root centre (with usually 3 or 4 points).
   - The vascular tissue forms a central cylinder through the root (stele).
   - The stele is surrounded by a pericycle.

7. The parenchyma (packing) cells store starch and other substances.

8. **Root hairs** increase the surface area for absorption.

9. Cap of cells protects the dividing cells behind the root cap, which are delicate and are easily damaged. The cap of cells also has a role in lubricating the root tip and facilitating root movement through the soil.

Leaf Structure (page 143)

1. (a) and (b), any two of the following:
   Hairs on the surface to reduce water loss, waxy cuticle to reduce water loss, sunken stoma to reduce water loss, modified leaves to store water (e.g. succulents), modification of leaves into tendrils to provide climbing support (e.g. pea plant), leaves are modified into spines for defence (e.g. cacti), leave are modified in carnivorous plants to catch prey.

2. The mesophyll is the main location for chloroplasts.

3. The air spaces in the leaf tissue increases the surface area allowing for a rapid exchange of gases (carbon dioxide in and oxygen out).

4. (a) Gas exchange occurs through small pores on the leaf surface called stomata. Carbon dioxide enters and oxygen and water exit.
   (b) Gas exchange is regulated by guard cells which surround each stoma. When the guard cell is filled with water (turgid) they swell, opening the stomatal pore. When water moves out of the guard cells, they lose turgor and become flaccid, causing the stomatal pore to close, preventing further exchange and water loss.

Xylem (page 144)

1. **Xylem** conducts water and dissolved minerals around the plant (from roots to leaves).

2. (a) Vessel elements: Provide rapid, low resistance conduction of water.
   (b) Tracheids: Role in water conduction but there is more impedence to water flow than in vessels.
   (c) Fibres (and sclereids): Provide mechanical support to the xylem.
   (d) Xylem parenchyma: Involved in storage.

3. Strengthened by having hard fibre cells and sclereids, and spiral thickening of the vessel walls.

4. As well as having pits in the walls (which tracheids also have) the end walls of vessels are perforated.

Phloem (page 145)

1. **Phloem** conducts dissolved sugar around the plant from its place of production to where it is required.
2. Xylem is dead, phloem is alive. Xylem transports water and minerals, phloem transports dissolved sugar.
3. Perforations enable the sugar solution to pass through and along the sieve tubes.
4. (a) The sieve cell or sieve tube member.
   (b) The companion cell keeps the sieve tube cell alive and controls its activity. They are responsible for loading and unloading sugar into the sieve cells.
5. Phloem parenchyma cells are involved in storage.
6. Strengthening: Fibres and sclereids.

Uptake at the Root (page 146)

1. (a) Passive absorption of minerals along with the water and active transport.
   (b) Apoplastic pathway (about 90%): Moving through the spaces within the cellulose cell wall. Symplastic pathway: Moving through the cell cytoplasm from cell to cell via plasmodesmata.
2. Large water uptake allows plants to take up sufficient quantities of minerals from the soil. These are often in very low concentration in the soil and low water uptakes would not provide adequate quantities.
3. (a) The **Casparian strip** represents a waterproof barrier to water flow through the apoplastic pathway into the stele. It forces the water to move into the cells (i.e. move via the symplastic route).
   (b) This feature enables the plant to better regulate its uptake of ions, i.e. take up ions selectively. The movement of ions through the apoplast cannot be regulated because the flow does not occur across any partially permeable membranes.

Transpiration (page 147)

1. (a) They take up water by the roots.
   (b) Any of:
   - Transpiration stream enables plants to absorb sufficient quantities of the minerals they need (the minerals are absorbed with the water and are often in low concentration in the soil).
   - Transpiration helps cool the plant.
2. Water moves by osmosis in all cases. In any order:
   (a) **Transpiration pull**: Photosynthesis and evaporative loss of water from leaf surfaces create a more negative water potential in the leaf cells than elsewhere in the plant, facilitating movement of water along a gradient in water potential towards the site of evaporation (stomata).
   (b) **Capillary effect/cohesion-adhesion**: Water molecules cling together and adhere to the xylem, creating an unbroken water column through the plant. The upward pull on the sap creates a tension that facilitates movement of water up the plant.
   (c) **Root pressure** provides a weak push effect for upward water movement.
3. (a)-(c) any of the following: High wind, high light, high temperature, low humidity. All increase the rate of evaporation from the leaves.
4. The system excludes air. As the plant loses water through transpiration, it takes up water from the flask via roots (or cut stem). The volume removed from the flask by the plant is withdrawn from the pipette; this can be measured on the pipette graduations.
5. (a) Measurements were taken at the start and at the end of the experiment in the same conditions (still air, light shade, 20°C). These rates should be the same (give or take experimental error). This indicates that the plant has not been damaged by the experiment and any results are therefore a real response to the experimental conditions.
   (b) Moving air and bright sunlight increase transpiration rate, because they increase the rate of evaporation from the leaves. **Note**: Lower humidity could also be said to increase transpiration rate (by increasing the gradient in water potential), but this would need to be tested further, i.e. the results here do not conclusively show this. Another test where the effects of darkness and humidity level were separated would be required. This is a good discussion point for students investigating experimental design and interpretation of results.
   (c) Still, humid conditions reduce evaporative loss, dark conditions stop photosynthetic production of sugars (therefore solute concentration in the leaves falls). Both these reduce transpiration rate by reducing the concentration gradient for water movement.

Adaptations of Xerophytes (page 149)

1. **Xeromorphic** adaptations allow xerophytes to survive and grow in areas with low or irregular water supplies.
2. (a)-(c), three in any order:
   - Modification of leaves to reduce transpirational loss (e.g. spines, curling, leaf hairs).
   - Shallow, but extensive fibrous root system to extend area from which water is taken and to take advantage of overnight condensation.
   - Water storage in stems or leaves.
   - Rounded, squat shape of plant body to reduce surface area for water loss.
3. The CAM metabolism (found only in xerophytic plants, many of which are succulents) allows carbon dioxide to be fixed during the dark. This produces organic acids which accumulate in the leaves and later release carbon dioxide into the Calvin cycle during daylight (when light energy is available to provide H^+ and ATP for photosynthesis). The stomata can then stay closed during the day when transpirational losses are highest.
4. A moist microenvironment reduces the gradient in water potential between the leaf and the air, so there is less tendency for water to leave the plant.
5. In a high salt environment, free water is scarce. Sea shoreline plants (halophytes) therefore have many xerophytic adaptations.

Translocation (page 151)

1. (a) '**Source to sink**' means the sugar flows from its site of production (in the leaves) to its site of unloading (at the roots).

(b) Usual source: Leaves and sometimes stems. Usual sink: Roots.
(c) Other source: Tubers or other storage organs from which sugar may be mobilised when photosynthetic tissues are absent.
(d) Other sink: Fruits where sugar is required to form the succulent tissues of the fruit.

2. The energy is required to generate the gradient in H^+ that is used to drive the transport of sucrose into the transfer cell.

3. (a) Translocation: The transport (around the plant) of the organic products of photosynthesis.
(b) The bulk movement of phloem sap along a gradient in hydrostatic pressure (generated osmotically).
(c) The coupling of sucrose transport (into the transfer cell) to the diffusion of H^+ down a concentration gradient (generated by a proton pump).

4. The increase in dissolved sugar in the sieve tube cell decreases its water potential (makes it more negative). Because of this, water moves into the sieve tube cells by osmosis (water moves to regions of lower water potential).

5. The transfer cell uses active transport mechanisms (coupled transport of sucrose) to accumulate sucrose to levels 2-3 times those in the mesophyll. The sucrose then moves into the sieve tube cell.

6. Xylem sap is only water and dissolved minerals; phloem sap is a 30% sugar (mainly sucrose) solution.

7. Transport of sugars in the phloem is active and requires energy to be expended. For this the tissue must be alive. Movement of water in xylem is a passive process.

8. If sap moved by pressure-flow, then there should be selective pressure for the sieve plate to be lost or become less of a barrier, yet this has not happened. (Of course, there are also selective pressures that operate against loss of the sieve plate, e.g. the need to have discrete yet freely communicating cells).

The Biochemical Nature of the Cell (page 155)

1. (a) Low viscosity: Water flows through small spaces and capillaries. It also enables aquatic organisms to move through it without expending a lot of energy.
(b) Colourless and transparent: Light penetrates tissue and aquatic environments. This property allows photosynthesis to continue at considerable depth.
(c) Universal solvent: It is the medium for the chemical reactions of life. Water is also the main transport medium in organisms.
(d) Ice is less dense than water: Ice floats and also insulates the underlying water.

2. (a) Lipids are important as a ready store of concentrated energy (their energy yield per gram is twice that of carbohydrates). They also provide insulation and a medium in which to transport fat-soluble vitamins. Phospholipids are a major component of cellular membranes.
(b) Carbohydrates are a major component of most plant cells, a ready source of energy, and they are involved in cellular recognition. They can also be changed into fats.
(c) Proteins are required for growth and repair of cells. They may be structural, catalytic, or have a variety of other functions as well as being able to be converted into fats.
(d) Nucleic acids, e.g. DNA and RNA, encode the genetic information for the construction and functioning of an organism.

Organic Molecules (page 156)

1. Carbon, hydrogen, and oxygen.

2. Sulfur and nitrogen.

3. Four covalent bonds (valency of 4).

4. A molecular (or chemical) formula shows the numbers and kinds of atoms in a molecule whereas a structural formula is the graphical representation of the molecular structure showing how the atoms are arranged.

5. A functional group is an atom or group of atoms, such as a carboxyl group, that replaces hydrogen in an organic compound and that defines the structure of a family of compounds and determines the properties of the family.

6. It is an aldehyde.

7. Either of: amine group (NH_2) or carboxyl group (COOH).

8. The amino acid cysteine has an R group (SH) that can form disulfide bridges with other cysteines to create cross linkages in a polypeptide chain (protein).

Biochemical Tests (page 157)

1. R_f = 15 mm ÷ 33 mm = **0.45**

2. R_f must always be less than one because the substance cannot move further than the solvent front.

3. Chromatography would be an appropriate technique if the sample was very small or when the substance of interest contains a mixture of several different compounds and neither is predominant.

4. Immersion would just wash out the substance into solution instead of separating the components out behind a solvent front.

5. Leucine, arginine, alanine, glycine (most soluble to least soluble).

6. Lipids are insoluble in water. They will not form an emulsion in water unless they have first been dissolved in ethanol (a non-polar solvent).

Water and Inorganic Ions (page 158)

1.

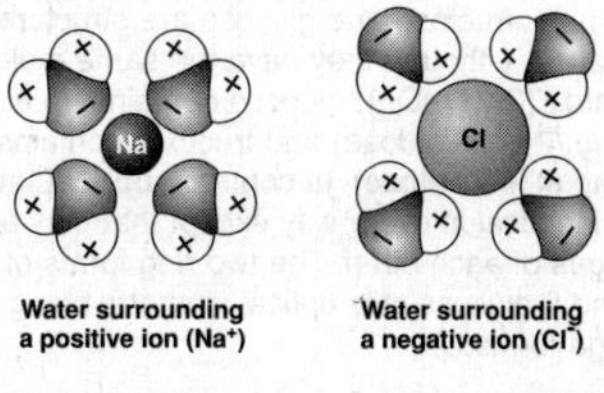

Water surrounding a positive ion (Na^+) | Water surrounding a negative ion (Cl^-)

2. The **dipole nature** of water means that it is a good solvent for many substances, e.g. ionic solids and other

polar molecules such as sugars and amino acids. It is therefore readily involved in biochemical reactions.

3. Inorganic compounds can be formally defined with reference to what they are not, i.e. organic compounds. Organic compounds are those which contain carbon, with the exception of a few types of carbon containing inorganic compounds such as carbonates, carbon oxides, and cyanides, as well as elemental carbon.

4. (a) Calcium: Calcium ions (Ca^{2+}) are a component of bones and teeth. Ca^{2+} also functions as a biological messenger.
Deficiency: Depletion of bone stores and increased tendency to bone fracture, disturbance to calcium regulating mechanisms, impairment of nerve and muscle function.
(b) Iron: Iron ions (Fe^{2+}) are a component of haemoglobin, the main oxygen carrying molecule, where Fe^{2+} is the central ion of the molecule.
Deficiency: Anaemia, fatigue, pallor, irritability, general weakness and breathlessness.
(c) Phosphorus: Phosphate ions (PO_4^{3-}) are a component of adenosine triphosphate (ATP), an energy-currency molecule which stores energy in an accessible form. Bone is calcium phosphate.
Deficiency: Anorexia, impaired growth, skeletal demineralisation, muscle atrophy and weakness, cardiac arrhythmia, respiratory insufficiency, decreased blood function, nervous system disorders, and even death.
(d) Sodium: Sodium ions (Na^+) have a role similar to potassium ions in the sodium-potassium pump.
Deficiency: Electrolyte disturbances and water intoxication (toxic water levels in the blood).
(e) Sulfur: As part of four amino acids, sulfur is important in a number of the redox reactions of respiration, in carbohydrate metabolism, in protein synthesis, liver function, and in blood clotting. Hydrogen sulphide (H_2S) replaces H_2O in photosynthesis of some bacteria.
Deficiency: Rare, but sulfur deficiency is attributed to circulatory problems, skin disorders, and various muscle and skeletal dysfunctions.
(f) Nitrogen: Nitrogen is a constituent element of all living tissues and amino acids.
Deficiency: Notable in plants with stunting of growth and yellowing of leaves. In animals, nitrogen deficiency manifests as various types of protein deficiency disease, e.g. kwashiorkor, which is characterised by degeneration of the liver, severe anaemia, oedema, and inflammation of the skin.

Carbohydrates (page 159)

1. **Structural isomers** have the same molecular formula but their atoms are linked in different sequences. For example, fructose and glucose are structural isomers because, although they have the same molecular formula ($C_6H_{12}O_6$), glucose contains an aldehyde group (it is an aldose) and fructose contains a keto group (it is a ketose). In contrast, **optical isomers** are identical in every way except that they are mirror images of each other. The two ring forms of glucose, α and β glucose, are optical isomers, being two mirror image forms.

2. Isomers will have different bonding properties and will form different disaccharides and macromolecules depending on the isomer involved, e.g. glucose and fructose are structural isomers; glucose + glucose forms maltose, glucose + fructose from sucrose. A polysaccharide of the α isomer of glucose forms starch whereas the β isomer forms cellulose.

3. Compound sugars are formed and broken down by condensation and hydrolysis reactions respectively. **Condensation reactions** join two carbohydrate molecules by a glycosidic bond with the release of a water molecule. **Hydrolysis reactions** use water to split a carbohydrate molecule into two, where the water molecule is used to provide a hydrogen atom and a hydroxyl group.

4. Cellulose, starch, and glycogen are all polymers of glucose, but differ in form and function because of the optical isomer involved, the length of the polymers, and the degree of branching. **Cellulose** is an unbranched, long chain glucose polymer held by β-1,4 glycosidic bonds. The straight, tightly packed chains give cellulose high tensile strength and resistance to hydrolysis. **Starch** is a mixture of two polysaccharides: amylose (unbranched with α-1,4 glycosidic bonds) and amylopectin (branched with α-1,6 glycosidic bonds). The α-1,4 glycosidic bonds and more branched nature of starch account for its physical properties; starch is powdery and more easily hydrolysed than cellulose, which exists as tough microfibrils. **Glycogen**, like starch, is a branched polymer. It is similar to amylopectin, being composed of α-glucose molecules, but it is larger and more there are more α-1,6 links. This makes it highly branched, more soluble, and more easily hydrolysed than starch.

Lipids (page 161)

1. In **phospholipids**, one of the fatty acids is replaced with a phosphate; the molecule is ionised and the phosphate end is water soluble. **Triglycerides** are non-polar and not soluble in water.

2. (a) Solid fats: Saturated fatty acids.
(b) Oils: Unsaturated fatty acids.

3. The amphipathic nature of phospholipids (with a polar, hydrophilic end and a hydrophobic, fatty acid end) causes them to orientate in aqueous solutions so that the hydrophobic 'tails' point in together. Hence the bilayer nature of phospholipid membranes.

4. (a) Saturated fatty acids contain the maximum number of hydrogen atoms, whereas unsaturated fatty acids contain some double-bonded carbon atoms.
(b) Saturated fatty acids tend to produce lipids that are solid at room temperature, whereas lipids that contain a high proportion of unsaturated fatty acids tend to be liquid at room temperature.
(c) The cellular membranes of an Arctic fish could be expected to contain a higher proportion of unsaturated fatty acids than those of a tropical fish species. This would help them to remain fluid at low temperatures.

5. (a) and (b) any of the following:
- Male and female sex hormones (testosterone, progesterone, oestrogen): regulate reproductive physiology and sexual development.
- Cortisol: glucocorticoid required for normal carbohydrate metabolism and response to stress.

- Aldosterone: acts on the kidney to regulate salt (sodium and potassium) balance.
- Cholesterol is a sterol lipid and, while not a steroid itself, it is a precursor to several steroid hormones and a component of membranes.

6. (a) Energy: Fats provide a compact, easily stored source of energy. Energy yield per gram on oxidation is twice that of carbohydrate.
   (b) Water: Metabolism of lipids releases water. **Note**: oxidation of triglycerides releases twice as much water as carbohydrate.
   (c) Insulation: Heat does not dissipate easily through fat therefore thick fat insulates against heat loss.

Amino Acids (page 163)

1. Comprise the building blocks for constructing proteins (which have diverse structural and metabolic functions). Amino acids are also the precursors of many important molecules (e.g. neurotransmitters and hormones).
2. The side chains (R groups) differ in their chemical structure (and therefore their chemical effect).
3. Translation of the genetic code. Genetic instructions from the chromosomes (genes on the DNA) determine the order in which amino acids are joined together.
4. **Essential amino acids** cannot be manufactured by the human body, they must be included in the food we eat.
5. **Condensation reactions** involve the joining of two amino acids (or an amino acid to a dipeptide or polypeptide) by a peptide bond with the release of a water molecule.
6. **Hydrolysis** involves the splitting of a dipeptide (or the splitting of an amino acid from a polypeptide) where the peptide bond is broken and a water molecule is used to provide a hydrogen atom and a hydroxyl group.
7. The L-form.

Proteins (page 165)

1. (a) **Structural**: Proteins form an important component of connective tissues and epidermal structures: collagen, keratin (hair, horn etc.). Proteins are also found scattered on, in, and through cell membranes, but tend to have a regulatory role in this instance. Proteins are also important in maintaining a tightly coiled structure in a condensed chromosome.
   (b) **Regulatory**: **Hormones** such as insulin, adrenaline (modified amino acid), glucagon (peptide) are chemical messengers released from glands to trigger a response in a target tissue. They help maintain homeostasis. **Enzymes** regulate metabolic processes in cells.
   (c) **Contractile**: Actin and myosin are structural components of muscle fibres. Using a ratchet system, these two proteins move past each other when energy is supplied.
   (d) **Immunological**: Gamma globulins are blood proteins that act as antibodies, targeting antigens (foreign substances and microbes) for immobilisation and destruction.
   (e) **Transport**: Haemoglobin and myoglobin are proteins that act as carrier molecules for transporting oxygen in the bloodstream of vertebrates. Invertebrates usually have some other type of oxygen carrying molecule in the blood.
   (f) **Catalytic**: Enzymes, e.g. amylase, lipase, lactase, trypsin, are involved in the chemical digestion of food. A vast variety of other enzymes are involved in just about every metabolic process in organisms.
2. Denaturation destroys protein function because it involves an irreversible change in the precise tertiary or quaternary structure that confers biological activity. For example, a denatured enzyme protein may not have its reactive sites properly aligned, and will be prevented from attracting the substrate molecule.
3. Any one of:
   - Globular proteins have a tertiary structure that produces a globular or spherical shape. Fibrous proteins have a tertiary structure that produces long chains or sheets, often with many cross-linkages.
   - The structure of fibrous proteins makes them insoluble in water. The spherical nature of globular proteins makes them water soluble.
4. (a) 21 amino acids (b) 29 amino acids

Enzymes (page 167)

1. Catalysts cause reactions to occur more readily. Enzymes are biological molecules (usually proteins) and allow reactions that would not otherwise take place to proceed, or they speed up a reaction that takes place only slowly. Hence the term, **biological catalyst**. The **active site** is critical to this function, as it is the region where substrate molecules are drawn in and positioned in such a way as to promote the reaction.
2. **Catabolism** involves metabolic reactions that break large molecules into smaller ones. Such reactions include digestion and cellular respiration. They release energy and are therefore **exergonic**. In contrast, **anabolism** involves metabolic reactions that build larger molecules from smaller ones. Anabolic reactions include protein synthesis and photosynthesis. They require the input of energy and are **endergonic**.
3. The **lock and key model** proposed that the substrate was simply drawn into a closely matching cleft (active site) on the enzyme. In this model, the enzyme's active site was a somewhat passive recipient of the substrate.
4. The **induced fit model** is a modified version of lock and key, where the substrate fits into the active site, and this initiates a change in the shape of the enzyme's active site so that the reaction can proceed.
5. (a) and (b) in any order, any two of:
   - Deviations from the optimum pH.
   - Excessively high temperature (heating).
   - Treatment with heavy metal ions, urea, organic solvents, or detergents.

   All these agents denature proteins by disrupting the non-covalent bonds maintaining the protein's functional secondary and tertiary structure. The covalent bonds providing the primary structure often remain intact but the protein loses solubility and the functional shape of the protein (its active site) is lost.
6. A mutation could result in a different amino acid being positioned in the polypeptide chain. The final protein may be folded incorrectly (incorrect tertiary and

quaternary structure) and lose its biological function. **Note**: If the mutation is silent or in a non-critical region of the enzyme, biological function may not be affected.

Enzyme Reaction Rates (page 169)

1. (a) An increase in enzyme concentration increases reaction rate.
   (b) By manufacturing more or less (increasing or decreasing the rate of protein synthesis).
2. (a) An increase in **substrate concentration** increases reaction rate to a point. Reaction rate does not continue increasing but levels off as the amount of substrate continues to increase.
   (b) The reaction rate changes because after a certain substrate level the enzymes are fully saturated by substrate and the rate cannot increase any more.
3. (a) An optimum **temperature** for an enzyme is the temperature at which enzyme activity is maximum.
   (b) Most enzymes perform poorly at low temperatures because chemical reactions occur slowly or not all at low temperatures (enzyme activity will reappear when the temperature increases; usually enzymes are not damaged by moderately low temperatures).
4. (a) Optimum **pH**: Pepsin: 1-2, trypsin: approx. 7.5-8.2, urease: approx. 6.5-7.0
   (b) The stomach is an acidic environment which is the ideal pH for pepsin.

Enzyme Cofactors and Inhibitors (page 170)

1. **Cofactors** are non-protein molecules or ions that are required for proper functioning of an enzyme either by altering the shape of the enzyme to complete the active site or by making the active site more reactive (improving the substrate-enzyme fit).
2. (a) Arsenic, lead, mercury, cadmium.
   (b) Heavy metals are toxic because they bind to the active sites of enzymes and permanently inactivate them. While the active site is occupied by the heavy metal the enzyme is non-functional. Because they are lost exceedingly slowly from the body, anything other than a low level of these metals is toxic.
3. (a) Examples: nerve gases, cyanide, DDT, parathion, pyrethrins (insecticides).
   (b) **Nerve gases** poison acetylcholinesterase, which is an important enzyme in the functioning of nerves and muscles (it normally deactivates acetylcholine in synapses and prevents continued over-response of nerve and muscle cells).
   Cyanide poisons the enzyme cytochrome oxidase, one of the enzymes in the electron transport system. It therefore stops cellular respiration.
   DDT and other organochlorines: Inhibitors of key enzymes in the nervous system.
   Pyrethrins: Insecticides which inactivate enzymes at the synapses of invertebrates. This has a similar over-excitation effect as nerve gases in mammals.
4. In **competitive inhibition**, the inhibitor competes with the substrate for the enzyme's active site and, once in place, prevents substrate binding. A **noncompetitive inhibitor** does not occupy the active site but binds to some other part of the enzyme, making it less able to perform its function as an effective biological catalyst.
5. Whilst noncompetitive inhibitors reduce the activity of the enzyme and slow down the reaction rate, **allosteric inhibitors** block the active site altogether and prevent its functioning completely.

DNA Molecules (page 172)

1. (a) 95 times more base pairs
   (b) 630 times more base pairs
2. < 2% encodes proteins or structural RNA.
3. (a) and (b) in any order:
   (a) Much of the once considered 'junk DNA' has now been found to give rise to functional RNA molecules (many with regulatory functions).
   (b) Complex organisms contain much more of this non-protein-coding DNA which suggests that these sequences contain RNA-only 'hidden' genes that have been conserved through evolution and have a definite role in the development of the organism.

Eukaryote Chromosome Structure (page 173)

1. (a) **DNA**: A long, complex nucleic acid molecule found in the chromosomes of nearly all organisms (some viruses have RNA instead). Provides the genetic instructions (genes) for the production of proteins and other gene products (e.g. RNAs).
   (b) **Chromatin**: Chromosomal material consisting of DNA, RNA, and histone and non-histone proteins. The term is used in reference to chromosomes in the non-condensed state.
   (c) **Histone**: Simple proteins that bind to DNA and help it to coil up during cell division. Histones are also involved in regulating DNA function in some way.
   (d) **Centromere**: A bump or constriction along the length of a chromosome to which spindle fibres attach during cell division. The centromere binds two chromatids together.
   (e) **Chromatid**: One of a pair of duplicated chromosomes produced prior to cell division, joined at the centromere. The terms chromatid and chromosome distinguish duplicated chromosomes before and after division of the centromere.
2. The chromatin (DNA and associated proteins) combine to coil up the DNA into a “super coiled” arrangement. The coiling of the DNA occurs at several levels. The DNA molecule is wrapped around bead-like cores of (8) histone proteins (called nucleosomes), which are separated from each other by linker DNA sequences of about 50 bp. The histones (H1) are responsible for pulling nucleosomes together to form a 30 nm fibre. The chromatin fibre is then folded and wrapped so that it is held in a tight configuration. The different levels of coiling enables a huge amount of DNA to be packed, without tangling, into a very small space in a well organised, orderly fashion.

Nucleic Acids (page 175)

1. (a)-(e) See below (only half of the section of DNA illustrated in the workbook is shown here):

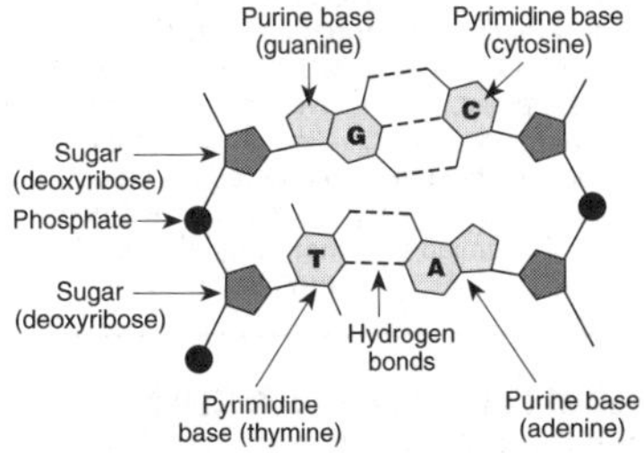

2. (a) The following bases always pair in a normal double strand of DNA: guanine with cytosine, cytosine with guanine, thymine with adenine, adenine with thymine.
   (b) In mRNA, uracil replaces thymine in pairing with adenine.
   (c) The hydrogen bonds in double stranded DNA hold the two DNA strands together.
3. **Nucleotides** are building blocks of nucleic acids (DNA, RNA). Their precise sequence provides the genetic blueprint for the organism.
4. The **template strand** of DNA is complementary to the **coding strand** and provides the template for the transcription of the mRNA molecule. The coding strand has the same nucleotide sequence as the mRNA (it carries the code), except that thymine in the coding strand substitutes for uracil in the mRNA.
5.

	DNA	RNA
Sugar present	Deoxyribose	Ribose
Bases present	Adenine	Adenine
	Guanine	Guanine
	Cytosine	Cytosine
	Thymine	Uracil
Number of strands	Two (double)	One (single)
Relative length	Long	Short

Creating a DNA Model (page 177)

3. Labels as follows:

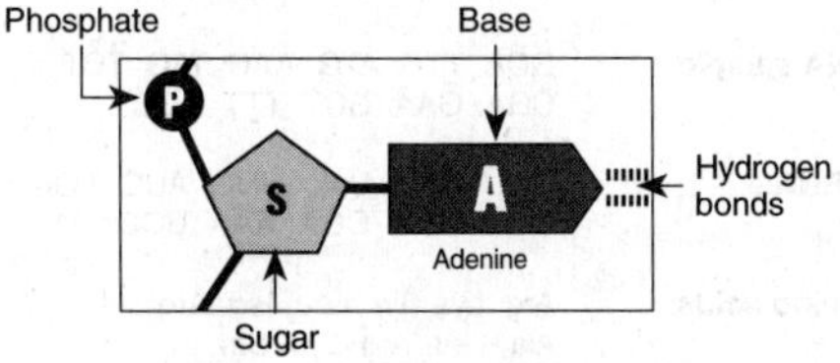

4. & 5. See the top of the next column.

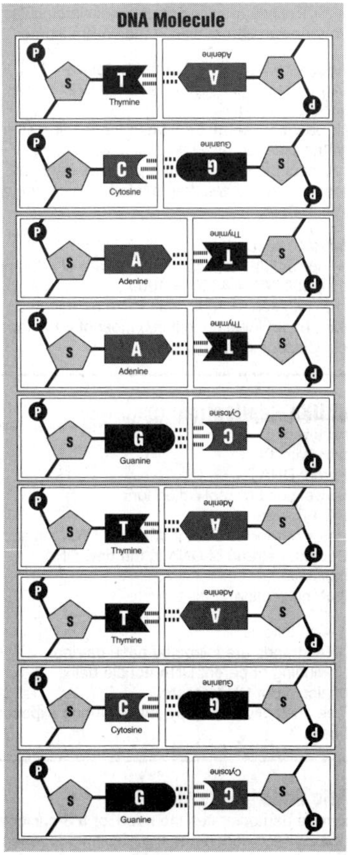

6. Factors that prevent a mismatch of nucleotides:
   - The number of hydrogen bond attraction points
   - The size (length) of the base (thymine and cytosine are short, adenine and guanine are long).

   Examples:
   Cytosine will not match cytosine because the bases are too far apart.
   Guanine will not match guanine because they are too long to fit side-by-side.
   Thymine will not match guanine because there is a mismatch in the number and orientation of hydrogen bonds.

DNA Replication (page 181)

1. DNA replication prepares a chromosome for cell division by producing two chromatids which are (or should be) identical copies of the genetic information for the chromosome.
2. (a) Step 1: Enzymes unwind DNA molecule to expose the two original strands.
   (b) Step 2: DNA polymerase enzyme uses the two original strands as templates to make complementary strands.

(c) Step 3: The two resulting double-helix molecules coil up to form two chromatids in the chromosome.

3. (a) **Helicase**: Unwinds the 'parental' strands.
   (b) **DNA polymerase I**: Hydrolyses the RNA primer and replaces it with DNA.
   (c) **DNA polymerase III**: Elongates the leading strand. It synthesises the new Okazaki fragment until it encounters the primer on the previous fragment.
   (d) **Ligase**: Joins Okazaki fragments into a continuous length of DNA.

4. 16 minutes 40 seconds
   4 million nucleotides replicated at the rate of 4000 per second: 4 000 000 ÷ 4000 = 1000 s
   Convert to minutes = 1000 ÷ 60 = 16.67 minutes
   (Note that, under ideal conditions, most of a bacteria's cell cycle is spent in cell division).

Review of DNA Replication (page 183)

1. (a) G pairs with C
   A pairs with T
   (b) A: Parent DNA
   B: Swivel point or replication fork
   C: DNA Polymerase III
   D: Parent strand of DNA in the new DNA molecule
   E: Daughter strand of DNA in the new DNA molecule
   F: Free nucleotide

3. Steps 1-4 as follows:
   1: DNA strands are joined by base pairing.
   2: Unwinding of parent DNA double helix.
   3: Unzipping of DNA parent.
   4: Free nucleotides occupy spaces along exposed bases.

The Genetic Code (page 184)

1. This exercise demonstrates the need for a 3-nucleotide sequence for each codon and the resulting degeneracy in the genetic code.

Amino acid	Codons	No.
Alanine	GCU GCC GCA GCG	4
Arginine	CGU CGC CGA CGG AGA AGG	6
Asparagine	AAU AAC	2
Aspartic Acid	GAU GAC	2
Cysteine	UGU UGC	2
Glutamine	CAA CAG	2
Glutamic Acid	GAA GAG	2
Glycine	GGU GGC GGA GGG	4
Histidine	CAU CAC	2
Isoleucine	AUU AUC AUA	3
Leucine	UAA UUG CUU CUC CUA CUG	6
Lysine	AAA AAG	2
Methionine	AUG	1
Phenylalanine	UUU UUC	2
Proline	CCU CCC CCA CCG	4
Serine	UCU UCC UCA UCG AGU AGC	6
Threonine	ACU ACC ACA ACG	4
Tryptophan	UGG	1
Tyrosine	UAU UAC	2
Valine	GUU GUC GUA GUG	4

2. (a) 16 amino acids
   (b) Two-base codons (eg. AT, GG, CG, TC, CA) do not give enough combinations with the 4-base alphabet (A, T, G and C) to code for the 20 amino acids.

3. Many of the codons for a single amino acid vary in the last base only. This would reduce the effect of point mutations, creating new and potentially harmful amino acid sequences in only some instances. **Note**: Only 61 codons are displayed above. The remaining three are **terminator** codons (labelled 'STOP' on the table in the workbook). These are considered the 'punctuation' or controlling codons that mark the end of a gene sequence. The amino acid **methionine** (AUG) is regarded as the 'start' (initiator) codon.

The Simplest Case: Genes to Proteins (page 185)

1. This exercise shows the way in which DNA codes for proteins. Nucleotide has no direct protein equivalent.
   (a) Triplet codes for amino acid.
   (b) Gene codes for polypeptide chain (may be a polypeptide, protein, or RNA product).
   (c) Transcription unit codes for functional protein.

2. (a) **Nucleotides** are made up of: Phosphate, sugar, and one of four bases (adenine, guanine, cytosine, and thymine or uracil).
   (b) **Triplet** is made up of three consecutive nucleotide bases that are read together as a code.
   (c) **Gene** comprises a sequence of triplets, starting with a start code and ending with a termination code.
   (d) **Transcription unit** is made up of two or more genes that together code for a functional protein.

3. Extra detail is provided; essential key words in bold. Steps in making a functional protein:
   - The template strand is made from the DNA coding strand and is transcribed into mRNA (**transcription**).
   - The code on the mRNA is translated into a sequence of amino acids (**translation**), which are linked with peptide bonds to form a polypeptide chain (this may be a functional protein in its own right).
   - The proteins coded by two or more genes come together to form the final functional protein (**folding into functional tertiary structure**).

Analysing a DNA Sample (page 186)

1. Use the *mRNA* table on the page: **The Genetic Code** in the workshop to determine the amino acid sequence.

Synthesised DNA	CGT AAG TAC TTG ATC AGA GCT CTT CGA AAA TCG
DNA sample:	GCA TTC ATG AAC TAG TCT CGA GAA GCT TTT AGC
mRNA:	CGU AAG UAC UUG AUC AGA GCU CUU CGA AAA UCG
Amino acids:	Arg Lys Tyr Leu Iso Arg Ala Leu Arg Lys Ser

2. (a) ATG ATC GGC GCT AAA TGT TAA
   (b) ATG CGG AAT TTC CCG GCT TAG
   (c) DNA replication

3. (a) mRNA: AUG AUC GGC GCU AAA UGU UAA
   Amino acids: Met Iso Gly Ala Lys Cys STOP
   (b) mRNA: AUG CGG AAU UUC CCG GCU UAG
   Amino acids: Met Arg Asn Phe Pro Ala STOP

(c) Protein synthesis

NOTE : In the next activities, including that on transcription, the **accepted convention** amongst molecular geneticists has been used, i.e. the **coding** (sense or non-transcribed) strand contains the same base sequence as the mRNA that is transcribed from the **template** (antisense) **strand**. This may oppose what appears in many texts, where there appears to be much confusion in the use of these terms. In fact, with the exception of template strand, the modern view is to avoid the use of these terms, as they imply that one strand alone carries the genes.

Transcription (page 187)

1. mRNA carries a copy of the genetic instructions from the DNA in the nucleus to ribosomes in the cytoplasm. The rate of protein synthesis can be increased by making many copies of identical mRNA from the same piece of DNA.
2. (a) AUG (b) UAA, UAG, UGA
3. (a) AUG AUC GGC GCU AAA
   (b) AUG UUC GGA UAU UUU

Translation (page 188)

1. AUG AUC GGC GCU AAA
2. (a) 61
   (b) There are 64 possible codons for mRNA, but three are terminator codons. 61 codons for mRNA require 61 tRNAs each with a complementary codon.

Review of Gene Expression (page 189)

1. (a) DNA
   (b) Free nucleotides
   (c) RNA polymerase enzyme
   (d) mRNA
   (e) Nuclear membrane
   (f) Nuclear pore
   (g) tRNA
   (h) Amino acids
   (i) Ribosome
   (j) Polypeptide chain
2. Process 1: Unwinding the DNA molecule.
   Process 2: mRNA synthesis: nucleotides added to the growing strand of messenger RNA molecule.
   Process 3: DNA rewinds into double helix structure.
   Process 4: mRNA moves through nuclear pore in nuclear membrane to the cytoplasm.
   Process 5: tRNA molecule brings in the correct amino acid to the ribosome.
   Process 6: Anti-codon on the tRNA matches with the correct codon on the mRNA and drops off the amino acid.
   Process 7: tRNA leaves the ribosome.
   Process 8: tRNA molecule is 'recharged' with another amino acid of the same type, ready to take part in protein synthesis.
3. Gene expression allows the cell to express the coded gene product (e.g. RNA or protein) only when needed.
4. Messenger RNA (mRNA), transfer RNA (tRNA), and ribosomal RNA (rRNA).
5. (a)-(c) any three of the following:
   DNA is double stranded, RNA is single stranded. DNA contains the sugar deoxyribose, RNA contains the sugar ribose. DNA contains the nucleic acids adenine, guanine, cytosine and *thymine*. RNA contains the nucleic acids adenine, guanine, cytosine and *uracil*. DNA carries all of the genetic information required by an organism to function and develop. RNA's are involved in 'reading' the genetic information and assembling the products it codes for.
6. (a) Transcription
   (b) Translation
7. Factors determining whether or not a protein is produced (in any order):
   (a) Whether or not the protein is required by the cell (regulated by control of gene expression).
   (b) Whether or not there is an adequate pool of the amino acids and tRNAs required for the particular protein in question.
8. DNA provides the genetic blueprint which codes for the sequence of amino acids from which proteins are built. DNA contain regulatory sequences which direct their synthesis, and can this way direct enzyme synthesis.
9. In **prokaryotic** gene expression, RNA is translated into protein almost as fast as it is transcribed; there is no nucleus therefore no separation of the transcription and translation processes. The presence of introns in a prokaryotic genome would interfere with protein function because there would be no time between transcription and translation to splice them out.
   In contrast, **eukaryotic** gene expression involves production of a primary RNA transcript from which the introns are removed. There is sufficient time for this to take place because transcription and translation occur inside and outside the nucleus respectively. **Note**: Evidence in support of this: Prokaryote DNA consists almost entirely of protein-coding genes and their regulatory sequences with very little non-protein coding DNA. Eukaryotic DNA comprises large amounts of non-protein coding sequences, much of which we now know codes for functional (regulatory) RNAs.

Global Human Nutrition (page 192)

1. In developing countries malnutrition is most commonly associated with undernutrition. Undernutrition describes a situation where the total food intake is not adequate. The range of foods consumed are often limited and deficient in certain nutrients causing malnutrition. In the developed world, the supply of food is plentiful and the amounts consumed often exceed the body's daily requirements. In this instance malnutrition arises from consuming an unbalanced diet which is calorie rich, highly processed, and contains high levels of sugar, fats and salts.
2. The modern industrialised agricultural techniques on which global food production is currently based are heavily reliant on fossil fuels, which are used to produce fertilisers and operate machinery. The decades following peak oil (when oil supplies peak and begins to decline), will see sharp increases in oil prices and consequently a rise in the costs of food production. This will causes food shortages through a number of interconnected mechanisms:
   – increased costs of food production will be

passed on as increased costs of food to consumers. This will disproportionately disadvantage poor nations where people will not be able to afford the increased costs.
- more of the world's food crops will be converted over to fuel production and there will be direct competition between fuel producers and food processors for supplies of wheat, corn, soybean, sugarcane, and other key crops.
- water short countries will divert water from irrigation and increase grain imports to feed their populations. This will reduce their internal (home-grown) supplies of food and make them even more susceptible to the vagaries of the global market.
- more countries with grain deficits.
- more frequent regional famines are a likely further consequence of all of the above.

A Balanced Diet (page 193)

1. Nutrients are required as (a) an energy source and (b) as the raw materials for metabolism, growth, and repair.

2. (a) The new food pyramid distinguishes between "healthy" and "unhealthy" fats, and "healthy" and "unhealthy" (refined) carbohydrate sources. This is contrary to the older pyramid, which placed all fats together as to be avoided, and recommended that all carbohydrates form the base of the food pyramid. The new pyramid recommends a high intake of plant oils which correlate with reduced rates of heart disease.
   (b) Evidence: Although heart disease rates are low in a country with low total fat intake (Japan), they are **even lower** in a country (Crete) where the total fat intake is very high, but the fats are mostly mono-unsaturates (from plant oils). Where saturated fat intakes are high, heart disease rates are also high.

3. Lactating women have a much higher requirements for calcium, protein, folate, and vitamin C than non-pregnant women.

4. (a) RDAs included recommended intakes for people with very high needs, so were often excessive in terms of the requirements of the average person.
   (b) RDAs recommended nutrient intakes for particular groups, but included those with very high needs. DRVs in contrast, provide guidelines based on a normal distribution of requirement in the population, where there is an estimated average requirement that will meet most people's needs.

5. (a) DRVs can be used to assess whether a diet is adequate in essential nutrients (and also energy). Diet can then be adjusted accordingly. **Note**: For those who must follow a low calorie diet, obtaining a nutrient balance is more difficult than with a high calorie diet, because intake is restricted.
   (b) DRVs on food labels allow consumers to more accurately assess food products and make more informed choices about what they add to their diets.

Deficiency Diseases (page 195)

1. **Malnutrition** refers to the inadequacy of the nutrition, whereas **starvation** refers to an absolute deficiency of calories (food in general of any sort).

2. (a) **Vitamin A**:
   Source: Animal liver, eggs, and dairy products
   Function: Required for the production of light-sensitive pigments in the retina and for the formation of cell structures.
   Deficiency: Loss of night vision, inflammation of the eye, keratomalacia and Bitots spots.
   (b) **Vitamin B_1**
   Source: Widely distributed in plant and animal food (meat, vegetables, wholemeal bread, and unpolished rice).
   Function: Required for respiratory metabolism.
   Deficiency disease: Beriberi, a form of nerve degeneration leading to weak muscles, impairment of growth, heart failure, and oedema.
   (c) **Vitamin C**:
   Source: Fresh fruit and vegetables.
   Function: Needed for production and metabolism of connective tissues, especially collagen.
   Deficiency: Poor wound healing, small blood vessel rupture, swollen gums and loose teeth.
   (d) **Vitamin D**:
   Source: Produced by the skin on exposure to sunlight.
   Function: Required for absorption of dietary calcium.
   Deficiency disease: In children, rickets, characterised by skeletal deformities, including typically bowed legs. In adults, osteomalacia produces similar symptoms.

3. Young children, pregnant women, and athletes have higher than normal energy requirements and some quite specific vitamin and mineral requirements. It is more difficult for these particular groups to satisfy their energy and nutrient requirements through normal eating. **Extra information**: **Children** are generally very active and are growing rapidly, so they have a high requirement for both energy and minerals such as iron and calcium. Children, with their smaller capacity, must eat more frequently. **Athletes** have high energy demands because of their increased physical activity and higher rates of tissue repair. Their need for some minerals, such as iron, is also higher than for non-active adults. Athletes must eat more, particularly of energy and nutrient dense foods. **Pregnant women** have slightly higher energy needs than non-pregnant women, as well as specific nutrient requirements (e.g. for iron and folate) associated with nourishing a developing foetus. Pregnant women should take care to avoid alcohol, and make food choices that supply their increased nutrient demands with only a moderate increase in calorie intake.

4. Iron is required for the production of haemoglobin and a lack of iron in the diet results in lower than normal levels of haemoglobin in red blood cells. Less oxygen will be able to be transported around the body and this results in breathlessness and fatigue (the typical symptoms of anaemia).

5. A zinc deficiency results in impaired muscle function because zinc is required for both enzyme activity and gene expression. Zinc is also important in sexual development, e.g. in the production of sperm, and a deficiency of zinc results in a delay in puberty.

6. Dietary supplementation can be achieved by adding the vitamin or mineral to a regularly used food product (e.g. folic acid and iron to breads and cereals, and iodine

to table salt). The benefits of dietary supplementation in this way stem from reaching a large sector of the general population easily in order to provide an essential element that may be deficient in the diet and contributes a lot to public health. In the case of iodine, which is widely deficient in soils and therefore also in food, supplementation is a public health exercise, which prevents goitre in the population (particularly poorer sectors who cannot afford to buy supplements).

7. Nutritional deficiencies result generally in poor system functioning; the skin, mucous membranes, and immune system will all be compromised when a person is malnourished. Such systems normally present a three tiered barrier to the invasion of pathogens. When they are not functioning, or functioning poorly, there is a greater susceptibility to disease.

Malnutrition and Obesity (page 197)

1. (a) Obesity is regarded as a form of malnutrition because more food is consumed than is required to stay healthy. The food consumed is often highly processed and not nutritionally balanced, containing high levels of fat, sugar, and salt.
   (b) The two basic energy factors which determine how a persons weight will change are the energy consumed and the energy expended (exercise).

2. (a) Overweight: 85.6-102.6 kg
   (b) Normal weight: 68.5-85.6 kg
   (c) Obese: over 102.6 kg

3. Diets high in saturated fat and sugar may cause atherosclerosis (fatty plaques which block arteries), which leads to (and is a symptom of) CVD. Poor diet is also implicated in high LDL:HDL ratios, obesity, and hypertension, which are risk factors for CVD.

4. (a) LDL deposits cholesterol on the endothelial lining of blood vessels, whereas HDL transports cholesterol to the liver where it is processed. A high LDL:HDL ratio is more likely to result in CVD because more cholesterol will be deposited on blood vessels and contribute to atherosclerosis.
   (b) The LDL:HDL ratio is a more accurate predictor of heart disease risk than total cholesterol per se.
   (c) The ratio can be lowered in at risk individuals by increasing exercise levels, stopping smoking, losing weight, decreasing consumption of saturated fat, and increasing intake of dietary fibre.

Cardiovascular Disease (page 199)

1. As it develops, **atherosclerosis** leads to blockage and obstruction of blood flow in the affected artery. The region of heart muscle normally supplied by the affected artery dies, causing severe pain and irregularity in the heart beat. Damage to the heart muscle may be so severe, it may lead to heart failure.

2. The technology for diagnosing and treating **CVD** has greatly improved so the chance of survival after an infarction is much better. In addition, because people are more aware of the symptoms and causes of CVD, they can change their lifestyles appropriately.

3. (a) **England and Wales**:
   Females: 151 ÷ 483 = **0.31**
   Males: 332 ÷ 483 = **0.69**
   i.e. much greater proportion of males die from CVD.
   (b) CVD is a less prevalent cause of death in women (at least until menopause) because female hormones (especially oestrogen) offer some protection for the cardiovascular system.

4. CVD incidence tends to higher in westernised countries where a high risk diet is also associated with a high stress, sedentary lifestyle (lack of exercise, high stress, faster pace of life, more motorised transport etc.).

5. (a) Controllable risk factors for CVD are those that can be altered by changing diet or other lifestyle factors, or by controlling a physiological disease state (e.g. high blood pressure, or high blood cholesterol). Uncontrollable risk factors are those over which no control is possible i.e. genetic predisposition, sex, or age. Note that the impact of uncontrollable factors can be reduced by changing controllable factors.
   (b) Controllable risk factors often occur together because some tend to be causative for others, or at least always associated, e.g. obesity greatly increases the risk of high blood lipids and high blood pressure: all factors increase the risk of CVD.
   (c) Those with several risk factors have a higher chance of developing CVD because the risks are cumulative and add up to pose a greater total risk.

6. Answers (a) and (b) for each factor below. All of these factors act to increase (directly or indirectly) the workload of the heart. The problems are usually not mutually exclusive, but related in their cause or effect.
   High blood pressure
   (a) Drives fat into the artery walls, encouraging atherosclerosis. The heart must work harder to counteract the resistance.
   (b) Keep blood pressure within normal limits by reducing other (mitigating) risk factors and/or by the use of controlling drugs.
   Cigarette smoking
   (a) Nicotine increases heart rate, blood vessel constriction, and blood pressure, causing the heart to work harder (nicotine operates via stimulation of the hormones aldosterone and adrenaline). At the same time, the amount of oxygen carried in the blood is reduced.
   (b) Give up smoking. Low tar, low nicotine cigarettes do not lower the risk of CVD (and may raise it).
   High blood cholesterol
   (a) Note: Applies really to high LDL (low density lipoprotein) which carries cholesterol and transports it to cells. Promotes accumulation of cholesterol and other fats in the artery walls, leading to atherosclerosis and artery obstruction (see question 1).
   (b) Keep a check on blood lipids (particularly the HDL:LDL ratio which ideally should be high). Reduce dietary intake of saturated fats, increase vigorous exercise (which raises HDL), lose weight.
   Obesity
   (a) Increases workload on the heart, which must pump blood through a larger network of blood vessels (created to supply a larger body). Associated with increased blood lipids, type II diabetes, and increased blood pressure.
   (b) Reduce weight through regular exercise and dietary management.

Type 2 diabetes
(a) People with poorly managed diabetes tend to have

many related complications associated with poor circulation that increase heart workload (see high blood cholesterol, high blood pressure).
(b) Manage diabetes through weight reduction, dietary management, and antidiabetic drugs.

High achiever/environmental stress
(a) These two factors are often (but not always) associated. High achievers and stressed people tend to over-respond to almost any type of stress. The release of adrenaline and the fight or flight response put increased workload on the heart and raise blood pressure (see high blood pressure).
(b) Reduce stress levels though better time management, regular exercise, recreation time etc.

Sedentary lifestyle
(a) Lack of physical activity leads to poor physical fitness, increased body fat, lowered capacity of the oxygen transport system, and inability to deal appropriately with stress (see obesity, stress). Muscles are not fit to work, leading to increased workload on the heart when activity is attempted.
(b) Increase physical activity by regular vigorous exercise (at least 3 times per week) .

The Green Revolution (page 201)

1. The first green revolution made use of increased use of machinery, and chemical fertilisers, pesticides, and water, as well as improvements in crop breeds in to increase the yield per unit land from key food crops. This greatly increased global food production but relied heavily on high energy inputs. The second green revolution has employed improved knowledge of plant breeding and genetics to produce high yielding strains with particular desirable qualities (such as disease resistance and drought tolerance) to increase yields further (particularly in tropical countries). Some of this second green revolution (especially more recently) has involved genetic engineering but some has been the result of improved selective breeding practices. This second green revolution is accelerating as a result of genome sequencing in crop varieties, although it is worth noting that continued gains in yield are frequently still reliant on high input agricultural techniques. The further step in the second green revolution is reduce reliance on chemical inputs.

2. (a) **Improve crop yields**:
Any crop improvements that reduce susceptibility to disease, allow easier crop management, or extend the cultivation range will increase yield. Examples include:
   - Cross breeding parental types to produce semi-dwarf cereal varieties with shorter stems than before. These produce better yields, as dwarf plants use more of their energy for filling the grain than in growing taller. Shorter plants are less likely to fall over, which also increases overall harvest yield (e.g. IR-8 rice)
   - Genetic engineering for disease and pest resistance, e.g. insertion of the Bt gene in corn, cotton, rice, and potatoes provides resistance to insect pests, resistance to powdery mildew in wheat.
   - Genetic engineering for environmental tolerance, e.g. salt tolerance in tobacco.
   - Genetic engineering for herbicide tolerance, e.g. glyphosphate tolerance in maize.

   (b) **Improve crop nutritional value:**
   Genetic engineering to increase micronutrient content of crops, e.g. provitamin A fortification in rice, increased protein content in potatoes, increased lysine content in maize.

3. (a) Recent crop developments could benefit countries suffering food shortages by providing access to high yielding, disease resistant varieties that are tolerant of a wide range of growing conditions. Potentially, new crop varieties could provide more grain, more cheaply, to more people without requiring more land to be under cultivation.
   (b) Developing countries might be unable to take advantage of these improvements if they are farming under sub-optimal conditions (e.g. lacking adequate rainfall or adequate irrigation and without access to manures or the money to purchase inorganic fertilisers.

Selective Breeding in Crop Plants (page 203)

1. Improving the phenotypic character of cereal crops was achieved by selecting the seed from the highest yielding and most pest/disease resistant plants from each year's crop and using only this seed to produce the crop for the following year. In a systematic way, over generations, the qualities of the stock would gradually move towards the desired phenotype. Over many years, the crop phenotype was altered so that it no longer so closely resembled the wild type from which it arose.

2. (a) Cauliflower: flowers
   (b) Kale: leaf
   (c) Broccoli: inflorescence
   (d) Brussels sprout: lateral buds
   (e) Cabbage: apical (terminal) bud
   (f) Kohlrabi: stem (swollen)

3. If allowed to flower, all six can cross-pollinate.

4. Unwanted plant species in our gardens (weeds) are selected against by a number of control methods. Physical weeding by hand or digging implement will favour those plants that have tough roots (e.g. dock) or propagation methods that are stimulated by weeding (e.g. oxalis). Heavy use of sprays will foster the development of herbicide resistance.

5. (a) and (b), any two of the following:
   - High yielding crops to maximise crop production.
   - Disease and pest resistant crops lead to increased crop yields and require less pesticide/herbicide application so maximise profit.
   - Fast growing varieties enable the crops to be harvested more quickly and allow more crops to be planted and harvested in a season.

Selective Breeding in Animals (page 205)

1. **Inbreeding** involves breeding between close relatives, and if practiced over a number of generations lead to increased homozygosity in a population. It is used by animal breeders to 'fix' desirable traits into a population, but an increase in the frequency of recessive, deleterious traits in homozygous form in a population can reduce the health and fitness and of a population and lower fertility levels. Out-crossing involves introducing new (unrelated) genetic material

into a breeding line. It is used to increase the genetic diversity, and is used in line-breeding to restore vigour and fertility to a breeding line.

2. Assisted reproductive technologies, such as artificial insemination, cryopreservation, embryo transfer and *in vitro* fertilisation, are used routinely to produce large numbers of offspring with desirable traits (e.g. high growth rates or superior wool production). These techniques allow the desirable traits to be fixed more quickly into the population than would be possible from traditional selective breeding techniques.

3. Positive outcomes of selective breeding in domestic animals include; desirable traits are established within a relatively short period of time, breeders are able to produce animals with high growth rates, animals can be selected for which produce high yields of meat, wool, or milk, production of animals with good temperament, improve the birthing characteristics of a breed (e.g. easy calving), produce animals specifically suited to the climate or terrain.

   Negative outcomes of selective breeding include; reduction of genetic variability can make the population susceptible to disease or physiological difficulties (e.g. hip displacement), fertility may decrease, the occurrence of deleterious genes becomes more widespread and the breed loses vigour.

4. Most genetic progress in dairy herds achieved by:
   (a) Selection of (and breeding from) high quality progeny from proven stock.
   (b) Extensive use of superior sires (breeding males) through artificial insemination.

5. **Genetic gain** refers to the gain towards a (reliably attained) desirable phenotype in a breed.

6. Mixed breeds combine the best of the characteristics of both species, i.e. optimum beef and milk production.

Producing Food with Microorganisms (page 207)

1. **Industrial microbiology** refers to the use of microorganisms, on an industrial scale, for the production of valuable commodities or to carry out some valuable process.

2. Production of alcoholic beverages:
   - Fermentation of fruit sugars by yeasts to produce wine.
   - Grains are first malted and the sugars fermented by yeasts to produce beer.

3. (a) Advantages: it is manufactured from recombinant microorganisms and not extracted from the stomachs of young calves, can be produced in large amounts relatively cheaply, produces a better flavour of cheese than microbial chymosin, cheese produced using GM chymosin is both vegetarian and kosher so can appeal to a wider market.
   (b) Disadvantages: some consumers perceive food made using GM ingredients as unsafe, chymosin is only one component of calf rennet, so other enzymes may need to be added during cheese making to mimic the 'natural' enzyme profile.

4. The general public often (wrongly) consider all bacteria to be harmful organisms capable of causing disease and illness, and so are wary of their use in food production. Yeast are considered a traditional ingredient as they have been used in the production of fermentation products (e.g. bread) for many thousands of years. Because of this they are more readily accepted by consumers.

Increasing Food Production (page 209)

1. **Advantages**: Increased crop growth rates and yields, production of a disease-free crop, the crop may be more aesthetically pleasing because it has no blemishes and fetch a premium price at the market, pesticides are fast acting and can control large infestations cost effectively,

   Disadvantages: Pesticides kill off all insects including beneficial ones, pesticides can accumulate in the environment and harm the local food chain (including humans), health risks to workers in the industry if pesticides are not handled correctly, pesticide resistance may develop from overuse or incorrect use, fertilisers may enter and contaminate waterways (leading to eutrophication), on a global scale fertiliser application emits greenhouse gases (N_2O) into the environment.

2. (a) Antibiotics are used in intensive farming systems to prevent, control and treat disease, allowing the production of healthy animals, which in turn allows for faster animal growth.
   (b) The use of antibiotics reduces the amount of disease occurring on intensive farming systems resulting in generally healthier animals overall (given the rearing system). In some cases, the prevention of animal disease may reduce the risk of the disease spreading to humans. Frequent use of antibiotics may result in the development of bacterial resistance, and there is some concern that this resistance may spread to humans. High density farming practices are considered by some to be inhumane and only viable because the use of antibiotics to keep disease at bay.

3. Sustainable agriculture is based on the principle of long term production with minimal environmental impact. It centres around farm management practices that ensure long term productivity and soil health.

4. (a) - (c) any three of the following:
   - Reduced use of fertilisers.
   - Reduced use of pesticides.
   - Smarter irrigation policies to conserve water.
   - Use of integrated pest management schemes.
   - Use of pulse crops and other natural sources for soil nitrification.

5. Sustainable agricultural practices can prove to be more profitable in the long term because they release farmers from the economic treadmill of having to apply increasing amounts of water, fertilisers, and pesticides to the land in order to achieve the same yields. With improved soil health, high energy, expensive inputs can be reduced to a minimum with no reduction in yield, thereby saving money and increasing profits.

Food Preservation (page 211)

1. (a) Heat treatments kill the majority of food spoilage organisms found in food, and by doing so, greatly increase its shelf life (deterioration or rotting of food is the result of microbial activity).
   (b) Heat treatments can sometimes alter the appearance and flavour of food. The degree of alteration is dependent upon the type of food being preserved and the heat treatment which is applied.

2. Sugar and salt both draw water away from the food by the process of osmosis. Without readily available water, the microorganisms on the food cannot grow, the bacterial population dies, and food spoilage is slowed.

Targets for Defence (page 214)

1. The natural population of (normally non-pathogenic) microbes can benefit the host by preventing overgrowth of pathogens (through competitive exclusion).

2. (a) The MHC is a cluster of tightly linked genes on chromosome 6 in humans. The genes code for MHC antigens that are attached to the surfaces of all body cells and are used by the immune system to distinguish its own tissue from that which is foreign.
   (b) This self-recognition system allows the body to immediately identify foreign tissue, e.g. a pathogen, and mount an immune attack against it for the protection of the body's own tissues.

3. Self-recognition is undesirable:
   - **During pregnancy**. **Note**: Some features of the self-recognition system are disabled to enable growth (to full term) of what is essentially a foreign body.
   - **During tissue and organ grafts/transplants** from another human (allografting) or a non-human animal (xenografting). **Note**: Such grafts are usually for the purpose of replacing rather than repairing tissue (e.g. grafting to replace damaged heart valves). For these grafts, tissue-typing provides the closest match possible between recipient and donor. The self-recognition system must also be suppressed indefinitely by immunosuppressant drugs.

Blood Group Antigens (page 215)

1. See table below:

Blood type	Antigen	Antibody	Can donate blood to:	Can receive blood from:
A	A	anti-B	A, AB	A, O
B	B	anti-A	B, AB	B, O
AB	A + B	none	AB	A, B, AB, O
O	None	anti-A + anti-B	O, AB, A, B	O

2. (a) Blood typing could eliminate who the murderer could not be (i.e. exclude some blood types).
   (b) It is very unlikely that blood typing could establish definitively who the murderer was.
   (c) A DNA (genetic) profile.
   (d) Blood typing is not used forensically because there are too many people who share a common blood type. The pool of "suspects" would be too large.

3. Discovery of the basis of the ABO system allowed the possibility of safe transfusions and greatly improved survival and recovery after surgery or trauma.

Blood Clotting and Defence (page 216)

1. (a) Prevents bleeding and invasion of bacteria.
   (b) Aids in the maintenance of blood volume.

2. (a) Injury exposes collagen fibres to the blood.
   (b) Chemicals make the surrounding platelets sticky.
   (c) Clumping forms an immediate plug of platelets preventing blood loss.
   (d) Fibrin clot traps red blood cells and reinforces the seal against blood loss.

3. (a) Clotting factors catalyse the conversion of prothrombin to thrombin, the active enzyme that catalyses the production of fibrin.
   (b) If the clotting factors were present all the time, the clotting could not be contained and the blood would clot when it should not.

4. (a) and (b) provided below. The first is the obvious answer, but there are disorders associated with the absence of each of the twelve clotting factors:
   (a) Classic haemophilia.
   (b) Clotting factor VIII (anti-haemophiliac factor).

   (a) Haemophilia B (Christmas disease).
   (b) Clotting factor IX (Christmas factor).

The Body's Defences (page 217)

1. The **first line of defence** provides non-specific resistance by forming a physical barrier to the entry of pathogens. Chemical secretions from the skin, tears, and saliva also provide antimicrobial activity and help destroy pathogens and wash them away. The **second line of defence** provides non-specific resistance operating inside the body to inhibit or destroy pathogens (irrespective of what type of pathogen is involved). Whereas the **third line of defence** provides specific defence resistance against particular pathogens once they have been identified by the immune system (antibody production and cell-mediated immunity).

2. **Specific resistance** refers to defence against particular (identified) pathogens. It involves a range of specific responses to the pathogen concerned (antibody production and cell-mediated immunity). In contrast, **non-specific resistance** refers to defence against any type of pathogen.

3. The leucocytes involved in the second line of defence are the monocytes (which mature as macrophages) and the granulocytes (eosinophils, neutrophils, and basophils), so-called because of the granular appearance of their cytoplasm. Their roles are:
   - Eosinophils produce antimicrobial substances, including proteins toxic to certain parasites. Eosinophils also show some phagocytic properties.
   - Basophils release heparin (an anticoagulant) and histamine which is involved in inflammation and

allergic reactions.
- Neutrophils and macrophages are phagocytic, actively engulfing and destroying foreign material (e.g. bacteria).

4. Functional role for (b)–(i) as follows:
   (b) **Phagocytosis** destroys pathogens directly by engulfing them.
   (c) Sticky **mucus** traps pathogens and **cilia** move the trapped microbes towards the mouth and nostrils.
   (d) Some secretions (sebum) have a pH unfavourable to microbial growth. The pH of **gastric juice** is low enough to kill microbes directly. Other secretions (**tears**, **saliva**) wash microbes away, preventing them settling on surfaces, sweat contains an enzyme that destroys some types of bacterial cell walls, **urine** flushes microbes from the urinary tract.
   (e) Antimicrobial properties of some proteins (e.g. **interferon**) prevent multiplication of microbes (especially viruses).
   (f) Produced against specific pathogens, **antibodies** bind and destroy pathogens or their toxins.
   (g) **Fever** raises general body temperature and metabolic rate which speeds up the blood flow and the rate of delivery of white blood cells to the site of infection. Fever also intensifies the effect of interferon.
   (h) **T cells** recognise and destroy target pathogens on contact. Other T cells assist by regulating the activity of other lymphocytes.
   (i) Heat inhibits the activity of the pathogens at the site of infection. Swelling and pain help to confine the infection to a limited area by limiting movement, increased blood flow speeds up the delivery of white blood cells and speeds healing.

5. With few T cells, the body lacks an effective cell mediated immune system and responds poorly to opportunistic pathogens that get past the first defences.

The Action of Phagocytes (page 219)

1. Neutrophils, eosinophils, macrophages.
2. By looking at the ratio of white blood cells to red blood cells (not involved in the immune response). An elevated white blood cell count (specifically a high neutrophil count) indicates microbial infection.
3. Microbes may be able to produce toxins that kill phagocytes directly. Others can enter the phagocytes, completely filling them and preventing them functioning or remaining dormant and resuming activity later.

Inflammation (page 220)

1. (a) **Increased diameter and permeability of blood vessels**. Purpose: Increases blood flow and delivery of leucocytes to the area. Aids removal of destroyed microbes or their toxins. Allows defensive substances to leak into the tissue spaces.
   (b) **Phagocyte migration and phagocytosis**. Purpose: To directly attack and destroy invading microbes and foreign substances.
   (c) **Tissue repair.** Purpose: Replaces damaged cells and tissues, restoring the integrity of the area.
2. Phagocytic features: Ability to squeeze through capillary walls (amoeboid movement), and ability to engulf material by phagocytosis.
3. Histamines and prostaglandins attract phagocytes to the site of infection.
4. **Pus** is the accumulated debris of infection (dead phagocytes, damaged tissue, and fluid). It accumulates at the site of infection where the defence process is most active.

Fever (page 221)

1. The high body temperature associated with fever intensifies the action of interferon (a potent antiviral substance). Fever also increases metabolism, which is associated with increased blood flow. These changes increase the rate at which white blood cells are delivered to the site of infection and help to speed up the repair of tissues. The release of interleukin-1 during fever helps to increase the production of T cell lymphocytes and speeds up the immune response.
2. **1**: Macrophage ingests a microbe and destroys it.
   2: The release of endotoxins from the microbe induces the macrophage to produce interleukin-1 which is released into the blood.
   3: Interleukin-1 travels in the blood to the hypothalamus of the brain where it stimulates the production of large amounts of prostaglandins.
   4: Prostaglandins cause resetting of the thermostat to a higher temperature, causing fever.

The Lymphatic System (page 222)

1. **Lymph** has a similar composition to tissue fluid but has more leucocytes (derived from lymphoid tissues). **Note**: Tissue fluid is similar in composition to plasma (i.e. containing water, ions, urea, proteins, glucose etc.) but lacks the large proteins found in plasma.
2. Lymph returns tissue fluid to general circulation, and with the blood, circulates lymphocytes around the body.
3. (a) **Lymph nodes**: Filter foreign material from the lymph by trapping it in fibres. They also produce lymphocytes.
   (b) **Bone marrow**: Produces many kinds of white blood cells: monocytes, macrophages, neutrophils, eosinophils, basophils, T and B lymphocytes.

The Immune System (page 223)

1. (a) **Humoral immune system**: Production of antibodies against specific antigens. The antibodies disable circulating antigens.
   (b) **Cell-mediated immune system**: Involves the production of T cells which destroy pathogens or their toxins by direct contact or by producing substances that regulate the activity of other cells in the immune system.
2. In the bone marrow (adults) or liver (foetuses).
3. (a) Bone marrow (b) Thymus
4. (a) **Memory cells**: Retain an antigen memory. They can rapidly differentiate into antibody- producing plasma cells if they encounter the same antigen again.
   (b) **Plasma cells**: Secrete antibodies against antigens (very rapid rate of antibody production).

(c) **Helper T cells**: Activate cytotoxic T cells and other helper T cells. Also needed for B cell activation.
(d) **Suppressor T cells**: Regulate the immune system response by turning it off when antigens disappear.
(e) **Delayed hypersensitivity T cells**: Cause inflammation in allergic responses and are responsible for rejection of transplanted tissue.
(f) **Cytotoxic T cells**: Destroy target cells on contact (by binding and lysing cells).

5. **Immunological memory**: The result of the differentiation of B cells after the first exposure to an antigen. Those B cells that differentiate into long lived memory cells are present to react quickly and vigorously in the event of a second infection.

Antibodies (page 225)

1. **Antibodies** are proteins produced in response to antigens; they recognise and bind antigens. **Antigens** are foreign substances (often proteins) that promote the formation of antibodies (invoke an immune response).

2. (a) The immune system must be able to recognise self from non-self so that it can recognise foreign material (and destroy it) and its own tissue (and not destroy it).
(b) During development, any B cells that react to the body's own antigens are selectively destroyed. This process leads to self tolerance.
(c) Autoimmune disease (disorder).
(d) Any two of: Grave's disease (thyroid enlargement), rheumatoid arthritis (primarily joint inflammation), insulin-dependent diabetes mellitus (caused by immune destruction of the insulin-secreting cells in the pancreas), haemolytic anaemia (premature destruction of red blood cells), and probably multiple sclerosis (destruction of myelin around nerves).

3. Antibodies inactivate pathogens in four main ways: **Neutralisation** describes the way in which antibodies bind to viral binding sites and bacterial toxins and stop their activity. Antibodies may also **inactivate particulate antigens**, such as bacteria, by sticking them together in clumps. Soluble antigens may be bound by antibodies and fall out of solution (**precipitation**) so that they lose activity. Antibodies also activate **complement** (a defence system involving serum proteins), tagging foreign cells so that they can be recognised and destroyed.

4. (a) **Phagocytosis**: Antibodies promote the formation of inactive clumps of foreign material that can easily be engulfed and destroyed by a phagocytic cell.
(b) **Inflammation**: Antibodies are involved in activation of complement (the defence system involving serum proteins which participate in the inflammatory response and other immune system activities).
(c) **Bacterial cell lysis**: Antibodies are involved in tagging foreign cells for destruction and in the activation of complement (the defence system involving serum proteins which participate in the lysis of foreign cells).

Acquired Immunity (page 227)

1. (a) **Active immunity** is immunity that develops after the body has been exposed to a microbe or its toxins and an immune response has been invoked.
(b) **Naturally acquired** active immunity arises as a result of exposure to an antigen such as a pathogen, e.g. natural immunity to chickenpox. **Artificially acquired** active immunity arises as a result of vaccination, e.g. any childhood disease for which vaccinations are given: diphtheria, measles, mumps, polio etc.

2. (a) **Passive immunity** describes the immunity that develops after antibodies are transferred from one person to another. In this case, the recipient does not make the antibodies themselves.
(b) **Naturally acquired** passive immunity arises as a result of antibodies passing from the mother to the foetus/infant via the placenta/breast milk. **Artificially acquired** passive immunity arises as a result of injection with immune serums e.g. in antivenoms.

3. (a) Newborns need to be supplied with maternal antibodies because they have not yet had exposure to the everyday microbes in their environment and must be born with operational defence mechanisms.
(b) The antibody "supply" is (ideally) supplemented with antibodies in breast milk because it takes time for the infant's immune system to become fully functional. During this time, the supply of antibodies received during pregnancy will decline.

New Medicines (page 228)

1. Some 120 drugs commonly used today are derived from plants, many of them from just a few species of tropical plants. Given that there are some 300 000 plus species of plants on Earth, the potential to discover and harvest new medicines from plants is vast, and thus far unrealised.

2. The majority of plant based medicines to date have been extracted from tropical plants. The loss of tropical plant biodiversity may mean that plant species are lost before we have had an opportunity to screen them for potential new medicines, and limit the number of new medicine discoveries.

Vaccination (page 229)

1. (a) and (b) See the table at the top of the next page.

2. (a) 280 days
(b) Antibody levels gradually build to a small peak after 40 days, then gradually decline to very low levels.
(c) Antibody levels rise very rapidly to a peak (much higher than that achieved after the first injection). Levels then decline slowly over a long period of time.
(d) The immune system has been "primed" or prepared to respond to the antigen by the first exposure to it (this initial response took a considerable time). When the cells of the immune system receive a second exposure to the same antigen they can respond quickly with rapid production of antibodies.

3. (a) The benefits achieved from childhood immunisation are gained from the protection the children receive from many contagious and potentially life threatening diseases. It is suggested that proper immunisation schedules, if followed, will lead to the elimination of many harmful diseases (as with the eradication of smallpox). It is also suggested that exposure to pathogens in a weakened or inactive

Vaccination schedule available to children in the United Kingdom								
Vaccine	Diseases protected from	Age / months				Age / years		
		2	3	4	12-15	3-5	10-14	13-18
DTP (Triple antigen)	Diphtheria, tetanus, pertusis (whooping cough)	✓	✓	✓				
Hib vaccine*	Haemophilus influenzae type b infection*	✓	✓	✓				
OPV (Sabin vaccine)	Poliomyelitis	✓	✓	✓		✓		✓
MMR	Measles, mumps, and rubella (german measles)				✓	✓		
BCG	Tuberculosis						✓	
DT booster	Diphtheria and tetanus					✓		
Td booster	Tetanus, Diphtheria (low strength dose)							✓
Vaccination schedules are also available ***for high risk groups*** for the following diseases: anthrax, hepatitis A, hepatitis B, influenza, pneumococcal disease, typhoid, varicella (chickenpox), and yellow fever.								

* Leads to meningitis in 60% of cases. Other problems include severe respiratory infections and septicemia. Dependingon an individual's vaccine tolerance, the Hib vaccine may be conjugated with the DTP vaccine or given as a separate vaccination from helps to strengthen a growing immune system.

(b) There is concern about immunisation because of the potentially dangerous side effects that can occur (although these complications are rare). Side effects of concern are very high fevers and seizures (fits) which can lead to nervous system (esp. brain) damage. Those opposed to childhood immunisation also claim that the child's immature immune system is not equipped to deal with the sudden onslaught of different antigens and is adversely affected (perhaps to the extent that the ability to combat common diseases such as colds and flu is compromised).

4. (a) Asia especially India: Cholera (*Vibrio cholerae*). Note that because this vaccine does not offer reliable protection against different strains, some health professionals are reluctant to provide it.
   (b) Any regions with poor sanitation: Typhoid fever (killed *Salmonella typhi* or live, oral vaccine)
   (c) Tropical areas such as Central and South America and Africa: Yellow fever (attenuated live strain of yellow fever virus).

Types of Vaccine (page 231)

1. (a) **Whole agent vaccine**: Made using entire microorganisms which are killed or weakened and therefore made non-virulent.
   Examples: Vaccines against influenza, measles, mumps, rubella, poliomyelitis, whooping cough.
   (b) **Subunit vaccine**: Made using fragments of a microorganism or a product of the microorganism that is capable of causing an immune response.
   Examples: Vaccines against diphtheria, meningococcal meningitis, tetanus, *Haemophilus influenzae* type b, hepatitis B, whooping cough.
   (c) **Inactivated vaccine**: Made by killing viruses by treating them with formalin or other chemicals.
   Examples: Vaccines against influenza and poliomyelitis (Salk vaccine).
   (d) **Attenuated vaccine**: Made by weakening the virus. Usually this is done by long-term culturing until so many mutations accumulate that the virus becomes non-virulent.
   Examples: Vaccines against mumps, measles, rubella, poliomyelitis (Sabin vaccine).
   (e) **Recombinant vaccine**: Developed with genetic engineering techniques (involving the transfer of DNA between organisms). The gene for some antigenic property of the pathogen (e.g. protein coat) is isolated and spliced into the genome of a benign vector, or a bacterium or yeast which can be cultured to produce large quantities of the antigen.
   Examples: The new vaccines against smallpox and hepatitis B (see 4(b)).
   (f) **Toxoid vaccine**: Bacterial toxins are inactivated with heat or chemicals. The deactivated toxins (toxoids) retain their antigenic properties and can stimulate the production of antibodies when injected.
   Examples: Vaccines against tetanus and diphtheria.
   (g) **Conjugated vaccine**: Made by combined highly antigenic toxoids with parts of another pathogen that are poorly antigenic (e.g. polysaccharide capsules). This makes the less effective antigen more effective.
   Example: *Haemophilus influenzae* type b vaccine.
   (h) **Acellular vaccine**: Made by fragmenting a whole-agent vaccine and collecting the antigenic portions.
   Example: Vaccine against hepatitis B, newer whooping cough vaccines, typhoid vaccines.

2. **Attenuated viruses** are more effective in the long term because they tend to replicate in the body, and the original dose therefore increases over time. Such vaccines are derived from mutations accumulated over time in a laboratory culture, so there is always a risk that they will back-mutate to a virulent form.

3. Heat kills by denaturing proteins. If the viral proteins are denatured, the virus loses its antigenic properties.

4. (a) They (vaccines made using recombinant methods) are safer because the vaccine is antigenic but not pathogenic. **Note**: Antigenic properties of the pathogen are retained, but there is no risk of developing the disease (as with a live, attenuated vaccine). With recombinant vaccines, the antigen can be produced in large quantities at relatively low cost using large scale culture techniques.
   (b) See the diagrams at the top of the next page (**only one of the two methods is required**).

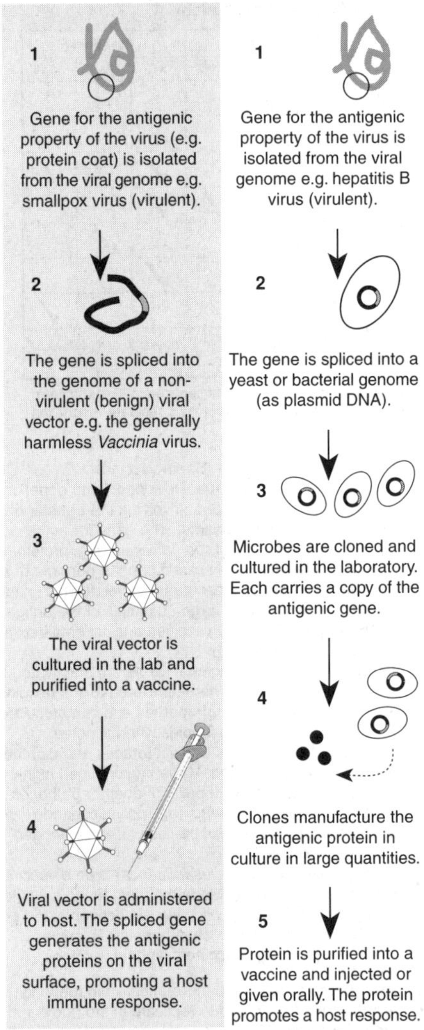

Health vs Disease (page 235)

1. **Disease** describes a condition where all or part of an organism's physiology is disturbed in some way (as a result of infection, an inherited disorder of metabolism, poor diet or lifestyle, or toxins). **Health** is not just the absence of disease, but a state of complete physical, mental, and social well-being.

2. Diseases are usually characterised by a suite of symptoms affecting a number of body systems, and there are many ways in which to view their classification. For example, mental disorders such as anorexia, manifest as deficiency diseases (starvation), and some so-called social diseases, such as STIs are the result of pathogens. Overlap in the causes of disease also leads to confusion over classification, e.g. there is a strong inherited predisposition to alcoholism (a social disease), as well as Alzheimer's and cancers (degenerative diseases).

Infection and Disease (page 236)

1. (a) A **pathogen** is an infectious disease causing agent, e.g. pathogenic bacterium, virus, fungus, prion etc.
   (b) There are many parasitic pathogens which can cause disease in humans. Some common organisms include *Cryptosporidium parvum, Giardia lamblia, Schistosoma, Toxoplasma gondii* and the hookworm. Although the effects of each parasitic pathogen are varied they all exploit the host for food, habitat and dispersal mechanisms and reduce the hosts fitness (health).

2. Robert Koch established a sequence of experimental steps, known as Koch's postulates, for directly relating a specific disease to its specific causative agent.

3. **Intracellular protozoan parasites** that cause disease are highly specialised with complex life cycles involving one or more hosts. Because they reside within the cells, it is difficult to control or treat them without also damaging the host (human) cells. In addition, many pass part of their life cycle as cysts, during which they are relatively resistant to treatment.

Transmission of Disease (page 237)

1. A parasite or pathogen benefits from entering a host's body because it offers a regular, easily accessible source of food, and a stable, relatively constant environment in which the parasite or pathogen can complete part or all of its life cycle.

2. (a) and (b) any order: Washing hands after going to the toilet or blowing nose. Sneezing or coughing into a tissue, which is then disposed of, rather than into a hand or into the air.

3. (a) **Malaria**: Transmission by animal (mosquito) vector. Portal of entry is a break in the skin surface. **Note**: This is an example of biological transmission where the vector is part of the pathogen's life cycle.
   (b) **Tetanus**: Transmission by indirect contact (e.g. as when injured with an infected rose thorn). Portal of entry is a break in the skin surface.
   (c) **Cholera**: Waterborne vehicle transmission. Portal of entry is the gastrointestinal tract.
   (d) **Common cold**: Direct contact and droplet transmission. Portal of entry is the respiratory tract.
   (e) **TB**: Droplet transmission. Portal of entry is the respiratory tract.
   (f) **HIV**: Direct contact transmission. Portal of entry are urinogenital openings and breaks in the skin surface.
   (g) **Gonorrhoea**: Direct contact transmission. Portal of entry are urinogenital openings.

Bacterial Diseases (page 238)

1. (a) - (b) any two of the following:
   - The use of toxins to break down the host cells or connective tissue which allow the bacteria to spread throughout the body.
   - The bacteria release enzymes to break down clot forming protein fibrin. This allows the bacteria to disperse more readily through the hosts blood system.
   - Fimbriae on the surface of the bacteria allow it to attach to mucous membranes and attack the hosts tissues.

2. (a) **Cholera** is spread by consumption of food or water contaminated with faecal material. The disease can be controlled by ensuring that all water and food consumed is clean and safe from faecal matter, and that good hygiene practices are followed to further reduce transmission.
   (b) **Salmonellosis** is most commonly contracted from eating raw or under-cooked poultry, egg and meat products contaminated with the bacterial genus *Salmonella*. The disease could be controlled by ensuring that all at risk foods were properly stored and cooked, and that safe food hygiene practices were followed to prevent cross contamination with other foods.

3. Airborne diseases are very easily spread when the infected person coughs or sneezes. Immunisation (vaccination) against airborne bacterial diseases would prevent a person becoming infected even if they did come in contact with a contagious person and help reduce the spread of the disease.

Tuberculosis (page 239)

1. The bacterium *Mycobacterium tuberculosis.*

2. The body's immune response is capable of walling-off the bacteria in nodules (tubercles). In this dormant state, the bacteria do not have the capacity to cause disease symptoms and the person is not infectious.

3. Air-borne transmission by coughing, sneezing, spitting (even talking at close range if saliva is transferred).

4. **Multi-drug resistance** arises when infected people fail to complete their course of treatment or when the course of treatment prescribed is inadequate. The bacteria are exposed to antibiotics for a time that is not long enough to eradicate them. When the drugs are withdrawn, subsequent generations arising from the surviving bacteria have the chance to develop drug resistance.

Foodborne Disease (page 240)

1. *E. coli* contamination through faecal contamination of food via unwashed hands, poor sanitation (inadequate sewage treatment and supply of clean water), faecal contamination of utensils or food preparation surfaces.

2. Developing countries tend to have poorer systems for the adequate treatment of sewage and provision of clean water.

3. Drinking water should be boiled or treated by filtration and chemicals (e.g. chlorine or iodine) to destroy microbes. Food should also be washed in treated water.

4. (a) **Salmonellosis**: Fever, accompanied by diarrhoea and abdominal cramps.
   (b) Transmission via contaminated foods of animal origin (especially poultry) following poor food handling and storage practices.

5. The bacteria will be destroyed by reheating, but not the toxins responsible for the poisoning. Staphylococcal exotoxins will only be destroyed after 30 minutes or so of boiling (and food is rarely reheated to this extent).

Cholera (page 241)

1. The bacterium *Vibrio cholerae.*

2. The symptoms of cholera include vomiting and acute, watery, painless diarrhoea. If untreated, the copious diarrhoea can quickly lead to severe dehydration and a consequent collapse of all body systems (particularly kidney and heart failure).

3. Transmission via contaminated food and water, particularly direct contact transmission spread as a result of inadequate hygiene or contact with contaminated foods or surfaces.

4. (a) Mild onset: Treated with oral rehydration.
   (b) Severe cases: Treated with intravenous fluids.

5. Risk factors associated with cholera include (primarily) unsanitary living conditions (including unsafe (unhygienic) disposal of faeces and faecal contamination of drinking water), and unhygienic food handling and preparation. These risk factors are prevalent in (economically) developing countries without the economic resources or infrastructure to provide proper housing, safe drinking water, and sewage treatment and disposal facilities.

Protozoan Diseases (page 242)

1. (a) A **cyst** is a protective capsule that allows survival of the protozoan when conditions are unfavourable.
   (b) The cyst enables a protozoan to survive because it protects the organism from adverse conditions, allowing the organism to survive outside a host and thus enabling it to reinfect another suitable host.

2. (a) Primary host = tsetse fly: Insecta: Diptera (flies), *Glossina* (several species, including *G. palpalis* and *G. morsitans*).
   (b) Primary host = *Anopheles* mosquito: Insecta: Diptera (flies), *Anopheles* (some 350 species).

3. Cysts of the giardia parasite are found in water contaminated with faecal material and often occurs in remote areas where reliable toileting facilities are not available (the parasite also favours conditions where the water is cool, as in forested streams).

4. Amoebic dysentery is most likely to be transmitted where there are a large number of infected individuals using one water supply for all their water requirements (hygiene, drinking), or where effluent is entering the drinking supply.

Malaria (page 243)

1. A mosquito carrying the plasmodium parasite bites a human. The mosquito injects a liquid (saliva) to prevent blood clotting that would block up its mouthparts. The parasite is injected along with the saliva.

2. By removing or destroying the mosquito breeding sites (stagnant water in which mosquito larvae live). This can be achieved by draining ponds and removing car tyres which are notorious for harbouring mosquito larvae. Scientists could assist by introducing mosquito eating fish, e.g. *Gambusia* and rotifers (although the efficacy of these as control agents is now in doubt and mosquito fish themselves are terrible pests).

3. (a) Headache, shaking, chills and fever, with coma, convulsions and death in extreme cases.
   (b) *Falciparum* malaria infects all ages of red blood cells whereas the other species attack only young or old cells. *Falciparum* infection results in the destruction of a greater number of cells.

4. The mosquito vectors require high (tropical) regions to breed. A warming climate would provide these conditions over a wider geographical range.

Resistance in Pathogens (page 244)

1. Factors contributing to the rapid spread of drug resistance in pathogens include:
   - The typically high mutation rates in viruses, bacteria, and protozoa, combined with short generation times (rapid generational turnover). Note: In bacteria and viruses, high mutation rates are the result of a higher error rate during DNA replication than is typical in most eukaryotic genomes.
   - Strong selection pressures imposed by drug use, coupled with misuse of drugs (e.g. using antibiotics against viral infections), poor patient compliance (patients not taking drugs as prescribed), and the poor quality of available drugs (especially to impoverished populations).

2. Answer will depend on student choice. All mechanisms are based on mutations that confer greater fitness in the prevailing environment:

- For bacteria: Genes for drug resistance (often carried on plasmid DNA) arise through mutation. In an environment that selects against susceptible strains, resistant bacteria will survive and increase in numbers. Genes for drug resistance are also easily transferred between strains, leading to a spread in resistance while that selection pressure remains. Drug resistance in bacteria can be conferred via a number of mechanisms, including inactivation of the drug, alteration of the drug's target, or alteration in the permeability of the bacterial cell to the drug.
- For HIV: Resistance may arise though a single mutation or through the accumulation of specific mutations over time. Such mutations may alter the binding capacity of the drug or susceptibility of the virus to the drug. Alternatively, resistance may arise as a result of naturally occurring polymorphisms, which gain favour in the selective environment.
- Chloroquine resistance in *Plasmodium falciparum* is based on the fact that they accumulate significantly less chloroquine than susceptible parasites. The mechanism for this appears to be due to a mutation conferring an enhanced ability to release the chloroquine from the vesicles in which it normally accumulates in the cell.

3. Health authorities can target multiple drug resistance by administering several drugs at the same time (what is called a drug cocktail). This does two things: it targets different stages of the pathogen's life cycle and it minimises the number of pathogens in the body. This multi-pronged attack reduces the pathogen "load" and minimises its reproduction/replication rate, while also reducing the chance of further drug resistant mutations arising.

Viral Diseases (page 245)

1. (a) **HIV**
   Disease: Acquired Immune Deficiency Syndrome
   Natural reservoir: Infected humans
   Symptoms: Exhausted immune system results in cancers and a wide range of opportunistic infections by bacteria, viruses, fungi, and protozoa.
   (b) **Coronaviruses**
   Disease: Colds and upper respiratory infections, SARS
   Natural reservoir: Infected humans
   Symptoms: Common cold (15% of cases). Sneezing, excessive nasal secretions, congestion of the respiratory passages. With SARS, high fever, pneumonia and, in some cases, death.
   (c) ***Influenzavirus***
   Disease: "Flu" (influenza)
   Natural reservoir: Infected humans. Pigs and waterfowl are reservoirs where the viruses reside and mutate into new strains.
   Symptoms: Chills, fever, headache, muscular aches, followed by cold-like symptoms.
   (d) **Filoviruses**
   Disease: Haemorrhagic fever (bleeding), e.g. Marburg, Ebola
   Natural reservoir: Unknown animal
   Symptoms: Severe headaches and skin rashes, bleeding from nose and gums followed by more widespread internal bleeding and destruction of internal organs.

2. Host specificity is based on the receptor properties of the viral protein coat. In enveloped viruses, the receptor regions are often glycoprotein spikes which provide a binding site for attachment to the host cell. As the features of the viral capsid or envelope are specific to the receptors of the host cell, only a limited host cell range will be susceptible to infection.

3. Viruses, particularly some types, show high mutation rates and the capsid and envelope proteins typically show relatively frequent change as a result of mutation. Even small changes in the receptor regions of the capsid/envelope have the potential to change the specificity so that cells (usually of the same type) in another species become susceptible, e.g. HIV infects T lymphocytes in humans and arose through mutations to the receptors regions of the pre-existing SIV (simian immunodeficiency virus) which also infects T cells.

4. Some viruses (e.g. *Influenzavirus*) acquire mutations (e.g. to their surface proteins) so rapidly that vaccines are difficult to develop and are quickly ineffective against new strains.

HIV and AIDS (page 247)

1. HIV attacks the very system that normally defends the body from infectious diseases. By knocking out the immune system, it leaves the body vulnerable to invasion by many microbes that would not normally infect a healthy person.
2. (a) The virus rapidly increases in numbers within the first year of infection, followed by a large drop off in numbers in the second year. Over the next 3-10 years, the HIV population gradually increases again.
   (b) The helper T cell numbers respond to the initial infection by increasing in numbers. After about a year, their numbers steadily decrease as they are attacked and destroyed by the HIV.
3. **Transmission of HIV**, any three of: Blood or blood products, vaginal secretions, breast milk, across the placenta, shared needles among intravenous drug users (contaminated with blood from other drug users), sexual intercourse: both homosexual (between males) and heterosexual (between men and women).
4. **HIV positive**: Blood tests have detected the presence of HIV in blood samples from a person (even though they may not have exhibited any symptoms).
5. Blood donated by the public and used to obtain a blood clotting factor (Factor VIII) for haemophiliacs was contaminated with HIV from donors already infected with the virus. This is particularly the case in countries where people are paid to donate blood.
6. HIV mutates very rapidly, changing its protein coat. Any vaccine is quickly rendered useless as the virus has already changed. There are also many different strains of HIV, each requiring a different vaccine.
7. Such people have a genetic makeup and a slightly altered metabolism that does not allow the HIV virus to replicate effectively inside their bodies. By determining how these people are different, it may be possible to develop drugs that mimic the altered metabolism.

Epidemiology of AIDS (page 249)

1. Men are likely to have **multiple partners concurrently**, and **avoid condom use**. **Women have little power** or say in controlling their circumstances and are at risk from partners engaging in risky sexual practices. The lack of employment in rural areas results in an **itinerant population of men**, moving between cities for work and developing sexual networks involving high risk behaviours. They contract HIV and return to spread it through their rural communities.
2. (a) Age structure: A high proportion of individuals of reproductive age are infected (and die) so that the population becomes dominated by non-reproductive individuals (old people and children). Young children, infected in the womb or during delivery, will also die.
   (b) Local economies will suffer through the removal (or incapacitation) of the sector of the population that normally provides most of the income. This group normally also supports children and elderly, so the economy also bears the cost of caring for the HIV infected, their children, and their old people.
3. Sub-Saharan Africa countries are underdeveloped in their technologies and economies and do not have the resources to fund an expensive and comprehensive AIDS treatment programme. Economicaly developed countries, which could support such drug programmes, choose not to.
4. **HIV-1**: Recombination of two existing strains of simian immunodeficiency virus (SIV) in a subspecies of common chimpanzee to produce a new strain. There was then cross species transmission of the virus to humans (zoonosis), probably through the handling and consumption of infected chimpanzee carcasses.
   HIV-2: Mutation of an existing simian immunodeficiency virus found in the sooty mangabey (SIV sm). Transmission as above.
5. See column graph below. **Note**: The axis break and change of scale made necessary by the large variability in cases per region and the space constraints here.

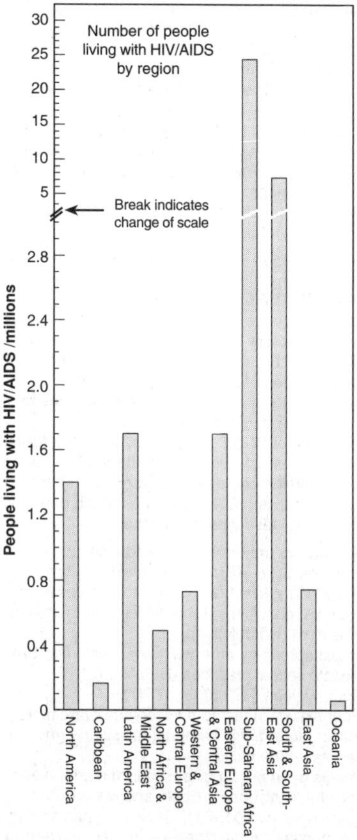

Replication in Animal Viruses (page 251)

1. Glycoprotein spikes are important in host recognition and attachment of the virus to the host cell.

2. (a) Endocytosis is the means by which foreign material is normally engulfed by cells, prior to being destroyed. This response of the cell enables the virus to gain entry into the cell.
   (b) Viral DNA replicated in the host cell's nucleus.
   (c) Viral proteins synthesised in the host cell's cytoplasm.

3. (a) HIV enters a cells by attaching to the CD4 receptors on a T cell, and fusing with the cell's plasma membrane.
   (b) Reverse transcriptase transcribes the viral RNA into viral DNA. This must occur for the viral genes to be able to integrate into the host's chromosomes where it stays as a provirus.
   (c) The provirus remains integrated with the host chromosome and persists as a latent infection. This means that it can reinfect new host cells whenever the DNA is replicated.

4. (a) Attachment: The virion comes into contact with a cell and adheres to receptor sites on the cell surface. The attachment structures on the viral surface match the receptors on the host cell.
   (b) Penetration: The host cell engulfs the attached viral particle by endocytosis.
   (c) Uncoating: The host's enzymes degrade the protein coat and release the viral nucleic acid into the cell.
   (d) Biosynthesis: The synthesis (assembly) of new infective virions.
   (e) Release: The active virions are budded off from the host cell by exocytosis.

Emerging Diseases (page 253)

1. Student's response depends on the disease chosen. **Biological factors** important in emergence and spread, which are worthy of discussion include:
   - Consumption of bush meat (SARS, HIV, haemorrhagic fevers), which allows viruses to cross the species barrier to humans.
   - Presence of insect or rodent vectors (*Hantavirus*, West Nile virus) that transmit the virus to humans.
   - Consumption of contaminated meat and poor stock feeding habits (vCJD); practices which have allowed crossing of diseases to humans.

   Social factors important in spread include:
   - Humans living in close proximity to vectors (*Hantavirus*, West Nile fever), to each other in crowded conditions (tuberculosis), or to animals in their care (avian influenza).
   - Rapid spread by air travel (most, including SARS).
   - Unsafe sexual practices (HIV/AIDS).
   - Poor hygiene (*E. coli* O157:H7).
   - Poor history of taking a full course of treatment, leading to a rise in bacterial resistance (multi-drug resistant tuberculosis).

   Examples of emerging diseases include: filoviral haemorrhagic diseases (Ebola and Marburg), *Hantavirus*, *E. coli* O157:H7, new influenza strains (e.g. avian influenza), prion diseases (vCJD), multi-drug resistant tuberculosis (re-emerging), HIV/AIDS, SARS.

2. A zoonosis is a disease arising in humans after crossing the species barrier from another animal (usually as a result of a mutation to the animal pathogen). Such diseases have the potential to cause highly lethal pandemics because they have no previous history in the human population (which will have only limited resistance to them). In addition, because they are new, there is often a considerable lag in identifying the cause and modes of transmission of zoonoses. This can make them more difficult to control quickly.

3. A **re-emerging disease**, e.g. TB, is one arising after the pathogen responsible develops resistance to the anti-microbial drugs used to control it. With renewed virulence, the pathogen can then spread rapidly.

4. Drug resistance has allowed the pathogens responsible for largely eradicated diseases, such as TB, to once again become established in the population. When cases appear they are much more difficult to treat and treatment is often ineffective, and this accelerates their spread through the population. In addition, vaccination programmes for some diseases may lapse so that when the disease is encountered again, a large proportion of the population will have no immunity to it.

5. **Biological factors** involved in the rapid spread of disease through hospitals include:
   - Close proximity of infected people to each other allows easy transference of infection between individuals, especially if hygiene practices are inadequate.
   - Compromised immune function of hospitalised individuals (people who are already unwell).
   - Ducted and linked air conditioning and heating systems transfer pathogens to many areas.
   - Food is prepared at the same place and disseminated to all patients.

   Social factors important in the spread include:
   - High concentration and close proximity of people with infections (also biological).
   - Movement of staff (who may carry infections) between patients, increasing the chances of transferring pathogens between individuals (also biological).

6. The death rate of Spanish flu was only 3%, so it was less virulent than the more recent SARS outbreak. However, it had a much greater global impact, killing millions, unlike SARS, which had little global impact biologically (although the economic impacts are potentially large and ongoing). Unlike the Spanish flu, the spread of SARS was limited early and managed well, with a global response to halt the disease.

7. At present, avian flu cannot pass directly from human to human and is not airborne. However, it is expected that, given time and a sufficiently high enough infection rate, it will mutate and be able do so. When (if) transmission becomes human to human and occurs by air or droplet infection, then global air travel provides the means to transfer the virus rapidly between populations. Precautions necessary to prevent spread presently include culling birds that have been positively identified as carrying the virus, banning the export of poultry from infected areas, and practising strict methods of hygiene in order to contain the disease. In the event of global spread, strict isolation, quarantine, and restrictions on air travel would be necessary.

8. Haemorrhagic fevers pose less risk of a pandemic than influenza because they are so virulent that people are debilitated by the disease before they move far. Outbreaks are then quickly isolated. In addition, humans are not the reservoir for the disease and the disease occurs in regions where the population is not

inclined, through socio-economic reasons, to travel far.

The Control of Disease (page 255)

1. Contagious diseases are those that are easily spread between people (or other animals), whereas non-communicable diseases are not spread between hosts.

2. (a) **Isolation** refers to the containment (in isolation facilities) of people with symptoms of a specific communicable disease so that they do not spread the disease to others. **Quarantine**, in contrast, is the physical isolation of those who are **potentially** infectious (i.e. may be infected) until the time of incubation for infection has passed.
   (b) Isolation and quarantine prevent the spread of a disease by preventing contact of infected or potentially infected people with uninfected individuals. In the case of SARS, immediate quarantine of individuals suspected (but not confirmed) of having the disease helped to contain its spread quickly and effectively.

3. Condoms help to prevent the spread of HIV by presenting a physical barrier to the spread of the virus between people who have sexual contact. HIV requires the transfer of body fluids (including semen).

4. The drainage of stagnant water helps to reduce the breeding areas for the Anopheles mosquito, which is the vector in the transmission of the malarial parasite.

5. (a) **Disinfectants** destroy microbes on non-living surfaces that might otherwise act as sites for transfer of microbes between hosts.
   (b) **Antiseptics** destroy microbes on living surfaces (e.g. skin) thus preventing transfer of microbes between hosts (antiseptics are kinder to skin than disinfectants).
   (c) **Heat** can be used to destroy microbes and/or their toxins, making utensils, food products etc. safe for use or ingestion (inadequate heating of food after storage is a common cause of food poisoning).
   (d) **Ionising radiation** sterilises by directly killing surface microbes. Materials for use (e.g. hospital dressings) must be packaged immediately to protect the sterile environment and prevent recontamination.
   (e) **Desiccation** inhibits the growth of microorganisms by removing water from the growing medium (e.g. food) and so presenting a dry environment in which growth occurs slowly or not at all. NOTE: Most bacteria cannot metabolise and grow when water is absent from the growth medium since water is required to utilise the available energy source. Desiccation can be achieved through the use of high sugar or salt solutions, which effectively reduce available water.
   (f) **Cold** inhibits microbial activity by reducing metabolic rates and therefore slowing growth. The lower the temperature, the slower the growth.

6. (a) Identification of the genetic cause of disease (e.g. the location of the genes associated with particular diseases and disorders) will enable rapid, effective genetic screening for particular diseases using DNA profiling techniques.
   (b) A knowledge of the location of the genes associated with specific disorders and diseases will facilitate the development of gene therapies for the correction of the faulty genes.

7. (a) A proportion of the population choose not to immunise against the disease; the vaccination programme is not 100% effective.
   (b) Vaccination creates a large pool of immune individuals in the population (herd immunity). With so few susceptible individuals, the measles virus is not easily transmitted through the population. **Note**: When the herd immunity effect fails as a result of a fall in vaccination rates, the population as a whole becomes more vulnerable to the disease.

Antimicrobial Drugs (page 257)

1. Ideally, an antimicrobial drug should have selective toxicity, targeting and killing the pathogen without harming the host. There is a wide range of antibiotics available; broad spectrum antibiotics, effective against a wide range of bacteria, are useful when the identity of the pathogen is unknown and a treatment decision must be made quickly. Narrow spectrum antibiotics are useful when the pathogen is known and can be targeted directly. The latter are the preferred choice as they limit the disturbance to the body's own microbial flora. Some patients exhibit side effects, ranging from discomfort to anaphylaxis, but the vast majority of people experience few difficulties with their use.

2. (a) **Antibiotic resistance** refers to the resistance some bacteria show to antibiotics that would normally inhibit their growth. In other words, they no longer show a reduction in growth response in the presence of the antibiotic.
   (b) It is important to finish a course of antibiotics so that the chances of survival of resistant mutants are minimised. If the course ends too quickly, more resistant strains may survive and flourish when antibiotic levels in the blood fall.

3. (a) **Advantages of broad spectrum drugs**: Useful when the identity of the pathogen is unknown, as valuable time can be saved in treating the infection. **Disadvantages of broad spectrum drugs**: The normal microbial flora of the body is also removed as well as the pathogen. This normal flora usually competes with and checks the growth of potential pathogens and other microbes. When it is removed, the body is exposed to infections from opportunistic pathogens, such as fungi.
   (b) Broad spectrum drugs: Tetracycline, streptomycin.

4. Antibiotics work by acting on the cellular materials or machinery of the pathogen (they inhibit some aspect of metabolism such as cell wall production or protein synthesis). This approach is of no use for viruses which are within the human host cells and are using the cellular machinery of the host cell to make viruses rather than normal cellular materials. The virus cannot be targeted without targeting the cell in which it resides.

5. Any four of:
   - Inhibit cell wall production, preventing cell division.
   - Inhibit gene copying (thereby preventing DNA replication and transcription).
   - Inhibit enzyme activity, preventing the synthesis of essential metabolites.
   - Damage the plasma membrane so that it ruptures.
   - Inhibit protein synthesis by interfering with translation.

6. Antibiotic A is most effective in controlling the growth of the bacterium because it has the most extensive zone around it where bacterial growth is prevented.

Diseases Caused by Smoking (page 259)

1. Long-term smoking results in increased production of mucus (in an attempt to trap and rid the lungs of smoke particles). This lung tissue is irritated and a cough develops associated with removing the excess mucus. The smoke particles indirectly destroy the alveolar walls, leading to coalescing of the alveoli and a substantial loss of lung surface area. The toxins in the smoke and tar damage the DNA of cells and lead to cancerous cells and tumours.

2. (a) **Tars**: Cause chronic irritation of the respiratory system and are also carcinogenic
   (b) **Nicotine**: Addictive component of tobacco smoke
   (c) **Carbon monoxide**: Markedly reduces the oxygen carrying capacity of the blood by binding to haemoglobin and forming a stable carboxy-haemoglobin compound. CO has a very high affinity for Hb (higher than that of oxygen) and will preferentially occupy oxygen binding sites. It is released only slowly for the body.

3. (a) **Emphysema**: Increasing shortness of breath (which becomes increasingly more severe until it is present even at rest). Chest becomes barrel shaped (associated with air being trapped in the outer part of the lungs). Often accompanied by a chronic cough and wheeze (caused by the distension (damage and coalescing) of the alveoli). **Note**: Chronic bronchitis and emphysema are often together called chronic obstructive lung disease.
   (b) **Chronic bronchitis**: A condition in which sputum (phlegm) is coughed up on most days during at least three consecutive months in at least two consecutive years. The disease results in widespread narrowing and obstruction of the airways in the lungs and often occurs with or contributes to emphysema.
   (c) **Lung cancer**: Impaired lung function; coughing up blood, chest pain, breathlessness, and death.

4. The evidence linking cigarette smoking to increased incidence of respiratory and cardiovascular diseases and cancer is clear and convincing. The incidence of these diseases in smokers is much higher than in non-smokers even when other factors such as age and familial tendency are considered. For example, deaths from lung cancer in the UK are five times higher in smokers than in non-smokers. These differences are large enough to be statistically significant. Importantly, the evidence linking smoking to the incidence of specific diseases comes from many different sources and is not conflicting.

The New Tree of Life (page 262)

1. The argument for the new classification as three domains is based on the fact that the genetic differences between the Bacteria and the Archaea are at least as great as between the Eukarya and the Bacteria. In other words, the traditional scheme does not accurately reflect the true evolutionary (genetic) relationship between the three groupings.

2. Any one of:
   - The eukaryote groups are given much less prominence, reflecting the true diversity of the prokaryote groups.
   - The Archaea have been separated out as distinct from other bacteria in order to reflect their uniqueness and indicate their true relationship to eukaryotes and to other prokaryotes.

3. The six kingdom classification scheme splits the prokaryotes into the kingdoms Eubacteria and Archaebacteria. These taxa are the same two domains that the three domain classification system uses.

New Classification Schemes (page 263)

1. (a) Morphology recognises the importance of physical features in distinguishing between groups of organisms (it is a simpler and more familiar operation). It also recognises the amount of morphological change that occurs in species after their divergence from a common ancestor.
   (b) Biochemical evidence produces phylogenies that more correctly represent the true evolutionary relationships between groups (taxa). **Note**: The phylogenies produced this way may be more difficult to interpret and apply and may not recognise morphological changes occurring after divergence from a common ancestor.

2. Biochemical evidence compares DNA and proteins between species and provides a more direct measure of common inheritance. **Teacher's note**: For some species, biochemical evidence has shown that earlier phylogenies were in error. Sometimes (as in the case of primates) the earlier phylogenies reflected the human view of their own position in the phylogeny. Morphological similarities can arise through convergent evolution in unrelated groups. Biochemical evidence is not clouded by this type of adaptive morphology.

3. (a) Pongidae (b) Hominidae

Classification System (page 269)

1. (a) 1. Kingdom, 2. Phylum, 3. Class, 4. Order, 5. Family, 6. Genus, 7. Species
   (b) 1. Animal, 2. Chordata, 3. Mammalia, 4. Primates, 5. Hominidae, 6. *Homo*, 7. *sapiens*

2. A two part naming system where the first word (capitalised and italicised) denotes the genus and the second word (lower case and italicised) denotes the species. Sometimes a third word (also lower case and italicised) denotes a subspecies.

3. (a) and (b) in any order:
   Avoid confusion over the use of common names for organisms; provide a unique name for each type of organism; attempt to determine/define the evolutionary relationship of organisms (phylogeny).

4. Any of the following:
   DNA profiling/sequencing: Where the unique genetic makeup of an individual is revealed and used for comparisons with related organisms.

DNA hybridisation: Where the percentage DNA similarity between organisms is compared.
Amino acid sequencing: Where the number of amino acid differences between organisms are compared.
Immunological distance: Indirectly estimate the degree of similarity of proteins in different species.

5. (a) **Monotreme**: Egg laying with little internal development before laying, most development takes place in the egg
   (b) **Marsupial**: Birth takes place after limited internal development. Most development occurs after 'foetus' moves to the pouch and attaches to the nipple.
   (c) **Placental**: Long period of internal development, sustained by placenta. Birth takes place at highly developed stage.

Features of the Five Kingdoms (page 271)

1. Distinguishing features of **Prokaryotae**: Lack nuclei and the organised chromosomes typical of eukaryotes. Some genetic material carried on plasmids. Small (70S) ribosomes but lack membrane-bound organelles. Most have a cell wall containing peptidoglycan (unique to bacteria). Divide by binary fission. Cell wall may be associated with a glycocalyx (capsule or slime layer). As a taxon, show a large diversity of nutritional modes and lifestyles.

2. Distinguishing features of **Protoctista**: Unicellular or simple multicellular eukaryotes. A diverse group of organisms, many of which are not related phylogenetically. Includes animal-like organisms such as protozoans and plant-like photosynthetic algae.

3. Distinguishing features of **Fungi**: Eukaryotic unicellular or multicellular organisms. Heterotrophic and lack chlorophyll. Saprophytes, parasites, or symbionts. Rigid cell walls of chitin. Nutrition is always absorptive. Typical organisational unit in filamentous forms is the hypha. Sexual/asexual reproduction involving spores.

4. Distinguishing features of **Plantae**: Multicellular eukaryotes, the large majority being photosynthetic autotrophs with chlorophyll. Clearly defined cellulose cell walls. Food stored as starch (and lipid). Primarily sexual reproduction with cycles of alternating haploid and diploid generations.

5. Distinguishing features of **Animalia**: Heterotrophic, multicellular eukaryotes. Cells lack a cell wall. Blastula stage in development. Further characterisation of animal taxa is based on body symmetry, type of body cavity (coelom), and internal and external morphology.

Features of Microbial Groups (page 272)

1. **Prokaryotae features**: Lack nuclei and the organised chromosomes typical of eukaryotes. Some genetic material carried on plasmids. Small (70S) ribosomes but lack membrane-bound organelles. Most have a cell wall containing peptidoglycan (unique to bacteria). Divide by binary fission. Cell wall may be associated with a glycocalyx (capsule or slime layer). As a taxon, show a large diversity of nutritional modes and lifestyles.

2. **Protoctista features**: Unicellular or simple multicellular eukaryotes. A diverse group of organisms, many of which are not related phylogenetically. Includes animal-like organisms such as protozoans and plant-like photosynthetic algae.

3. **Microfungi features**: Eukaryotic unicellular or microscopic multicellular organisms. Like all fungi, heterotrophic and lack chlorophyll. Nutrition is always absorptive. Saprophytes, parasites, or symbionts. Rigid cell walls of chitin. Typical organisational unit in filamentous forms is the hypha. Sexual/asexual reproduction involving spores.

Features of Animal Taxa (page 273)

Visible identifying features only including:

1. **Cnidarian features**: Body radially symmetrical. Medusoid (jellyfish) or hydroid (Hydra) body shape. Body wall of two layers separated by jellylike mesoglea. Tentacles with stinging cells for prey capture.

2. **Annelid features**: Body more or less cylindrical and obviously segmented. Move by peristalsis, short appendages, or whole body undulations (leeches).

General arthropod features: Insects, crustaceans, arachnids, and myriopods are all arthropods and share some common features as follows; an exoskeleton of chitin and protein that is shed periodically to allow growth. Paired jointed appendages (legs, mouthparts). Segmented body but with a tendency for loss, fusion or specialisation of segments to various degrees in different groups.

3. **Insect features**: Adult body composed of head, thorax and abdomen. Most adults have one or two pairs of wings (some have secondary loss of the wings). Juvenile forms may be maggot-like or similar to the adult. Three pairs of jointed legs off the thorax. All other appendages (including mouthparts) also paired and jointed. Compound eyes.

4. **Crustacean features**: Most are marine. Exoskeleton often impregnated with mineral salts (calcium carbonate and calcium phosphate). Body covered, at least partly, by a hard carapace. Paired, jointed appendages (often specialised to perform different functions). Two pairs of antennae. Gills often present. Compound or simple eyes. Diverse group; many are highly specialised.

5. **Arachnid features**: Exoskeleton of chitin and protein (as in other arthropods). Body divided into cephalothorax (combined head and thorax) and unsegmented abdomen. No antennae. First two pairs of appendages are feeding structures called chilicerae (fangs) and pedipalps. Terrestrial. Four pairs of walking legs. Many are poisonous with stings or venom sacs. Simple eyes. Spiders produce silk.

6. **Myriopod features**: Terrestrial. Elongated body with many obvious segments bearing paired legs. Head has a pair of antennae and two or three pairs of mouthparts. Centipedes have poison claws behind the mouthparts. Simple eyes. Most millipedes are slow moving, whereas many centipedes are rapid runners.

7. **Mollusc features**: Unsegmented. Muscular foot for movement. Upper body forms a fleshy mantle which may secrete a protective calcium carbonate shell. Scraping feeding organ or radula. Otherwise the group is very diverse. Aquatic and terrestrial. Commonly

recognised groups are the shelled gastropod snails and sea slugs and the bivalved molluscs (mussels etc.). The cephalopods (squids and octopus) are highly specialised: the foot is divided into tentacles and they have highly developed eyes.

8. **Echinoderm features**: Marine. Adults radially symmetrical. Body wall rigid (with a skeleton of tiny calcareous plates). Spines are common. Some (sea lilies) are attached, most are mobile. Fluid filled tube feet used for locomotion. In most, mouth points down.

Note: As follows - fish, amphibians, reptiles, birds and mammals are all vertebrates (internal bony skeleton). Paired nostrils and eyes.

9. **Fish features**: Aquatic, ectotherms. Skin slippery with mucus and covered with scales. Gills. Usually well developed fins and tail. Body shape varies depending on swimming mode. Obvious lateral line organ along body. Lack eyelids and tongue.

10. **Amphibian features**: Adaptations for both aquatic and terrestrial life. Ectothermic. Smooth (non-scaly) skin covered with mucous glands. Moveable eyelids. Tympanic membrane for ear present in frogs. Muscular, protusible tongue. Legs and feet (may be webbed) adapted both for swimming and for moving on land.

11. **Reptile features**: Largely terrestrial or semi-aquatic. Dry, thick skin with horny scales or plates. May have legs or be legless (snakes). Tympanic membrane often present. Most lay shelled eggs. Most have well developed eyes.

12. **Bird features**: Endotherms. Horny scales on feet , but feathers over most of the body. Beak, lacking teeth. Adapted for flight (wings, light skeleton) although some are secondarily flightless. Lay heavily shelled eggs.

13. **Mammal features**: Endotherms covered with hair or fur. Glandular skin. Young born live except monotremes which lay eggs. Mammary glands produce milk for young. Teeth often highly specialised. (Internal features: secondary palate and diaphragm.)

Features of Macrofungi and Plants (page 275)

1. **Macrofungi features**: Most are decomposers (saprophytic). Vary from single celled to large multicellular moulds. Rigid cell walls of chitin. Hyphae present in filamentous moulds. Spread by spores. Lichens are a symbiosis between a phototroph (usually an alga) and a fungus.

Note: Mosses, liverworts, ferns, gymnosperms, and angiosperms are all plants and are autotrophic.

2. **Moss and liverwort features**: Lack vascular tissues. Small and restricted to damp environments. Root-like rhizoids anchor plant in ground. In mosses, the plant body is leaf-like. In liverworts it may be leaf-like or a flattened thallus.

3. **Fern features**: True vascular plants. Plant body consists of underground stem or rhizome which bears the leaves (fronds) and roots. Conspicuous spore cases on the undersides of fronds. Clearly defined alternation of generations between large, leafy sporophyte and small, heart-shaped gametophyte.

4. **Gymnosperm features**: True vascular plants. Most are woody plants bearing seeds in cones. Other have naked seeds. Many are evergreen. Male cones shed pollen. Seeds are contained in the female cones.

Note: General angiosperm features: Vascular plants, producing flowers, fruits, and seeds.

5. **Monocot angiosperm features**: Herbaceous (non-woody). Flower parts in multiples of three. Leaves narrow with parallel veins. Seeds have a single cotyledon.

6. **Dicot angiosperm features**: Herbaceous or woody. Flower parts in multiples of four or five. Leaves often broader than in monocots with netted (branching) vein pattern. Seeds have two cotyledons.

The Classification of Life (page 276)

The classification used here recognises five kingdoms: **Prokaryotae** (bacteria, cyanobacteria), **Protoctista** (simple eukaryotes such as protozoans and algae), **Fungi** (mushrooms, moulds, yeasts), **Plantae** (multicellular green plants), and **Animalia** (multicellular animals). Viruses are not included in this classification; they are not regarded as living organisms and are difficult to classify using traditional taxonomic methods. Some researchers have suggested viruses comprise 'quasi-species' and as such may be considered living. Prions are also unable to be classified in this five kingdom system. They are infective particles, the true nature of which is not yet fully understood. Students should classify the organisms provided as follows:

PROKARYOTAE:
Staphylococcus bacteria, *Anabaena* cyanobacterium

PROTOCTISTA:
kelp seaweed, *Amoeba, Paramecium, Euglena*

FUNGI:
mushrooms, *Rhizopus* bread mould

PLANTAE:
Bryophyta: liverwort, moss

Filicinophyta: Boston lace fern

Angiospermophyta: **Monocot**: tulip; **Dicot**: African violets

Cycadophyta: cycad

Coniferophyta: pine tree

ANIMALIA:
Porifera (sponges): tube sponge

Cnidaria: *Hydra*, marine jellyfish

Platyhelminthes: liver fluke, tapeworm

Annelida: earthworm, polychaete worm

Mollusca: **Gastropoda**: garden snail; **Bivalvia**: scallop; **Cephalopoda:** squid

Echinodermata: heart urchin, sea star

Arthropoda: **Crustacea**: *Daphnia*, marine crab; **Chilopoda**: centipede; **Diplopoda**: millipede; **Arachnida**: tarantula, scorpion; **Insecta**: blowfly, flea, dragonfly

Chordata: **Chondrichthyes**: stingrey, shark; **Osteichthyes**: eel, seahorse

Amphibia: frog
Reptilia:
Squamata: python, lizard
Crocodilia: crocodile
Chelonia: tortoise
Aves: parrot
Mammals:
Prototheria: echidna
Metatheria: kangaroo
Eutheria: manatee, peccary

Classification Keys (page 282)

1. The case (presence or absence and specific features of the case).

2. A *Oxyethira*
B *Hudsonema*
C *Olinga*
D *Aoteapsyche*
E *Hydrobiosis*
F *Helicopsyche*
G *Triplectides*

3. **Insect order (common name)**
(a) Plecoptera (stoneflies)
(b) Hemiptera (bugs)
(c) Coleoptera (beetles)
(d) Odonata (dragonflies and damselflies)
(e) Lepidoptera (moths and butterflies)
(f) Trichoptera (caddisflies)
(g) Emphemoptera (mayflies)
(h) Megaloptera (dobsonflies)
(i) Diptera (true flies)

Keying Out Plant Species (page 284)

1. (a) Silver maple, *Acer saccharinum*
(b) Japanese maple, *Acer palmatum*
(c) Red maple, *Acer rubrum*
(d) Sugar maple, *Acer saccharum*
(e) Black maple, *Acer nigrum*

2. Any one of: The size of the tree or shrub, the colour of the bark and flowers, the shape of the winter buds and winged fruit.

3. Before a plant can be classified to species level a number of different features must be considered. One feature is often not sufficient to accurately distinguish between closely related species within the same genus.

4. The key must be for the genus of plant that the unidentified species belongs to as well as include the species of unidentified plant.

Global Biodiversity (page 286)

1. Species diversity refers to the number of different species within an area (species richness), while genetic diversity describes the diversity of genes within a particular species. Biodiversity is defined as the measure of all genes, species, and ecosystems in a region, so both genetic and species diversity are important in determining a region's total biodiversity.

2. Consideration of ecosystem diversity is very important when considering areas to set aside for conservation purposes because regions with diverse ecosystems will have higher levels of species richness, and the two measures are interdependent. A loss of habitat diversity will have a negative impact on the species richness of an area, and habitat diversity itself is contingent on species richness. Loss or decline in one will invariably result in loss or decline in the other.

3. The hotspot list is as follows:

1 **Tropical Andes**
The richest and most diverse hotspot where it is home to 20 000 endemic plants and at least 1500 endemic non-fish vertebrates.

2 **Sundaland**
Some of the largest islands in the world are found here in Southeast Asia. The second-richest hotspot in endemic plants, and well known for its mammalian fauna, which includes the orangutan.

3 **Mediterranean basin**
The site of many ancient and modern civilisations, it is the archetype and largest of the five Mediterranean-climate hotspots (also see nos. 9, 12, 19 and 22). One of the hotspots most heavily affected by human activity, it has 13 000 endemic plants, and is home to a number of interesting vertebrates such as the Spanish ibex.

4 **Madagascar and Indian Ocean islands**
Madagascar is a top conservation priority as this 'mini-continent' has undergone extensive deforestation. This hotspot is famous for reptiles such as chameleons and is home to all the world's lemur species.

5 **Indo-Burma**
An area stretching from the eastern slopes of the Himalayas through Burma and Thailand to Indochina. This region hosts the world's highest freshwater turtle diversity (43 species), and a diverse array of mammals. Several new ungulate species, such as the saola and giant muntjac, were recently discovered here.

6 **Caribbean**
One of the highest concentrations of species per unit area on Earth. Reptiles are particularly diverse (497 species are found here), 80 percent of which are found nowhere else. Non-fish vertebrates number 1518.

7 **Atlantic Forest region**
Once covering an area nearly three times the size of California, the Atlantic Forest has been reduced to about 7% of its original extent. It is most famous for 25 different kinds of primates, 20 of which are endemic. Among its best-known 'flagship species' are the critically endangered muriquis and lion tamarins.

8 **Philippines**
The most devastated of the hotspots, the forest cover has been reduced to 3% of its original extent. The Philippines is especially rich in endemic mammals and birds, such as the Philippine eagle.

9 **Cape Floristic Province**
This Mediterranean-type hotspot in southern Africa covers an area roughly the size of Ireland, and is now approximately 20% of its original extent. It is home to 8200 plant species, more than 5500 of which are endemic.

10 **Mesoamerica**
Forming a land bridge between two American continents, this hotspot features species representative of North and South America as well as its own unique biota. The spider and howler monkeys, Baird's tapir and

unusual horned guan are 'flagship species'.

11 **Brazilian Cerrado**
A vast area of savanna and dry forest, the Cerrado is Brazil's new agricultural frontier and has been greatly altered by human activity in the past few decades. Home to 4400 endemic plants and several well-known mammal species, including the giant anteater, Brazilian tapir, and maned wolf.

12 **Southwest Australia**
A Mediterranean-type system, this hotspot is rich in endemic plants, reptiles, and marsupials including numbat, honey possum and quokka. It is also home to some of the world's tallest trees, e.g. giant eucalyptus.

13 **Mountains of South-Central China**
An area of extreme topography, these mountains are home to several of the world's best-known mammals, including the giant panda, the red panda, and the golden monkey. This hotspot is largely unexplored and may hold many undiscovered species.

14 **Polynesia/Micronesia**
This hotspot comprises thousands of tiny islands scattered over the vast Pacific, from Fiji and Hawaii to Easter Island and is noteworthy for its land snails, birds, and reptiles. Hawaii has suffered some of the most severe extinctions in modern history, due in part to the introduction of non-native plant and animal species.

15 **New Caledonia**
One of the smallest hotspots yet it has the largest concentration of unique plants with five plant families found nowhere else on Earth. This hotspot also features many endemic birds, such as the kagu, a long-legged, flightless forest dweller representing an entire family.

16 **Choco-Darien Western Ecuador**
Some of the world's wettest rain forests are found here, and amphibians, plants and birds are particularly abundant. It has one of the highest levels of endemism of any hotspot with 210 endemic amphibian species of the 350 species found here.

17 **Guinean Forests of West Africa**
(in error, this hotspot was not numbered on the map). With the highest mammalian diversity of any hotspot, these forests are home to the rare pygmy hippopotamus and many other striking species, including the western chimpanzee, Diana monkey and several forest duikers. The numbers of these endemic mammals have been severely reduced by large-scale logging and hunting.

18 **Western Ghats/Sri Lanka**
The Western Ghats mountain chain and adjacent island of Sri Lanka harbour high concentrations of endemic reptiles; of 259 reptile species, 161 are found nowhere else on Earth. This hotspot is also home to a number of 'flagship species', including the lion-tailed macaque.

19 **California Floristic Province**
Extending along the coast of California and into Oregon and northwestern Baja California, Mexico, this is one of five hotspots featuring a Mediterranean-type climate of hot, dry summers and cool, wet winters. It is especially rich in plants, with more than 4000 plant species, almost half of which are endemic.

20 **Succulent Karoo**
The only arid hotspot, the Succulent Karoo of southern Africa is renowned for unique succulent plants, as well as lizards and tortoises. in Namaqualand, in the southern part of this hotspot, a seasonal burst of bloom in September attracts many tourists.

21 **New Zealand**
This hotspot claims a number of world-famous endemic bird species, including kiwi (a nocturnal, flightless bird), takahe (a diurnal, flightless bird), and the critically endangered kakapo (a large, flightless parrot).

22 **Central Chile**
This hotspot features an arid region as well as a more typical Mediterranean-type zone. Best known for its incredible variety of plant species but also features unusual fauna, including one of the largest birds in the Americas, the Andean condor.

23 **Caucasus**
Situated between the Black Sea and the Caspian Sea, Caucasus habitats range from temperate forests to grasslands. A diversity of plants have been recorded here with some 6300 species, more than 1600 of which are endemic.

24 **Wallacea**
Named for the 19th century naturalist Alfred Russel Wallace, this hotspot comprises the large Indonesian island of Sulawesi, the Moluccas and many smaller islands. The area is particularly rich in endemic mammals and birds.

25 **Eastern Arc Mountains/ Coastal Forests of Tanzania and Kenya**
A chain of upland and coastal forests, this hotspot claims one of the densest concentrations of endemic plant and primate species in the world. It is home to African violets and 4000 other plant species, as well as the 1500 remaining Kirk's red colobus monkeys.

Britain's Biodiversity (page 287)

1. Table and graph below:

	% of species		Key
Protozoa	22.8%	(82°)	
Algae	22.8%	(82°)	
Fungi	17.1%	(62°)	
Ferns & bryophytes	1.2%	(4°)	
Lichens	1.7%	(6°)	
Flowering plants	1.6%	(6°)	
Invertebrates	32.5%	(117°)	
Vertebrates	0.3%	(1°)	

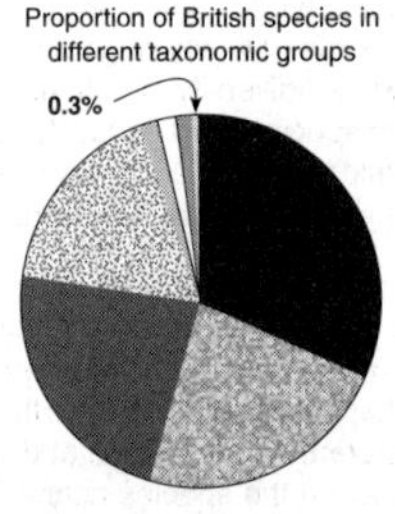

2. Invertebrate phyla, protozoans, algae, and lower plants show a much higher biodiversity (as measured by number of identified species) than the higher plants and vertebrate taxa.

3. (a) Our knowledge of the biodiversity (as measured by number of identified species) of invertebrate animals and bacteria (especially the latter) is poor compared with that of vertebrates i.e. there are many more undescribed species, and there is considerable doubt as to how many species actually exist.
   (b) Vertebrates are large, conspicuous organisms with (for the most part) sexual reproduction. This makes determination of "a species" relatively simple even if morphology is similar. **Note**: Many invertebrates and bacteria are as yet undescribed and there is much doubt as to what constitutes a species in those organisms that rarely reproduce sexually. This is especially the case for bacteria, where species classification tends to based upon criteria other than the ability to interbreed. For invertebrates that normally reproduce by parthenogenesis, identification of type specimens (specimens representing the species) is particularly difficult. Morphology may be very similar between different species and species distinction may be based only on physiological or genetic criteria.

4. The UK has a low level of endemism. It is not far from the European continent ; this facilitates species transfers. In addition, its isolation from the continental land mass has been limited for much of its geologic history (isolation is required for endemism to develop).

5. (a) 12 000 – 3800 = 8200
   8200 ÷ 12 000 X 100 = 68.3%. Nearly 70% decline.
   (b) The barn owl is widely distributed, so the population does not have the problems associated with scattered or isolated distribution. The causes of decline have been identifiable and manageable in that rather simple measures can be taken to increase survival rates of both the young and breeding adults. They are also relatively adaptable birds, taking a variety of prey and responding well to habitat enhancement (e.g. using new nest sites).

Loss of Biodiversity (page 289)

1. Loss of biodiversity from an ecosystem has a cascade effect to the remaining species. The effects depend very much on the species that disappears (e.g. predator, producer) but, in general, species loss results in altered food chains and food webs, allowing for the proliferation of some species and the demise of others. Other changes include a loss of stability and resilience, and disruption to normal processes, interactions and outcomes, such as nutrient cycling, soil formation, pollination, oxygen production, carbon sequestration and climate regulation.

Tropical Deforestation (page 290)

1. (a) They enhance removal of carbon dioxide from the atmosphere (anti-greenhouse).
   (b) They maintain species diversity.
   (c) They have, as-yet-undiscovered, potentially useful species for medicines etc.

2. Biodiversity hotspots are often located in tropical rainforests because they are very productive and nutrient rich environments cable of sustaining a large mass of life. Large numbers of species are able to flourish in these hotspots because of the diverse habitats it provides.

Biodiversity and Global Warming (page 291)

1. (a) Warmer climates will enable diseases and vectors for disease to spread into new regions.
   (b) Many annually occurring diseases are "killed off" in cooler months. Global warming may result in an extension of certain "disease seasons" if the temperature variation is insufficient to kill the organism.

2. Crop production would likely extend into the newly available arable land to meet increasing human food requirements. The biodiversity in these regions would be severely reduced as the native habitat was removed to make way for cropping.

3. Students discussion should include the following points:
   - Weather patterns will alter, and some regions will receive more rain than they do currently while others will receive less. Increases in temperature will extend the range for some crops and cause contraction in the growing range of others. There is likely to be a shift in traditional crop growing regions as a result.
   - Crops will be able to grow faster in the CO_2 enriched environment.
   - Food security will become more pressing if crops fail due to the changes caused by global warming.
   - Pests and disease will be able to move into new regions. This could have a negative impact of agriculture and human health in some regions.
   - Some species will be unable to adapt to the changing environment and will become widely or locally extinct in certain regions. Species loss will decrease the stability and resilience of ecosystems in affected areas.
   - Rising sea levels may result in salt water intrusion into aquifers and lead to increased salination of soils and lack of water for irrigation and drinking.
   - Increased frequency and severity of weather events will increase agricultural losses and reduce viability of food production in susceptible areas.
   - Shifts in agricultural ranges will exacerbate species losses in regions suited to food production (as land is turned over to agriculture).

Grassland Management (page 292)

1. (a) **Mowing**: Moderate mowing prevents the establishment of shrubs and woodland species and allows slower growing grassland species to compete with the "rough" grass species. This enhances the biodiversity of the grassland system.
   (b) **Burning**: Burning allows established moorland species to regrow vigorously from stem bases. Vigorous growth is needed if desirable species are to compete successfully with invasive weed species.
   (c) **Maintaining low soil fertility**: Low soil fertility promotes the growth of desirable, slower growing grassland species and discourages rapid growth of rough grass species.

2. (a) Advantage 1: High yields are maintained by intensive farming even though some areas are retired from production.
   (b) Advantage 2: There is a financial incentive if certain areas are left as conservation estate (income from tourism related to conservation areas and direct compensation for loss of income from land turned over to conservation).

3. Disadvantage: Intensive farming practices may have a lasting detrimental impact on surrounding conservation estate. The financial rewards of conserving land may not compensate for the income lost through having unproductive land.

The Impact of Farming (page 293)

1. Agricultural production involves the removal of potentially arable land from a natural "wilderness" state to one of (often) intensive management for crop or animal production. The conflict arises when land has a high value both for conservation estate and farm production. The decisions as to the fate of the land will depend on who the land belongs to and whether the conservation loss is acceptable if the land is put into production. Farmers with areas of high conservation value land on their own farm properties may be offered financial incentive not to develop the land. The availability of this type of compensation depends on the conservation value of the land, and the amount of compensation will be determined by how much the farmer stands to lose by not farming the land.

2. (a) and (b) any advantages of:
   - Hedgerows provide habitat and food for wildlife.
   - Hedgerows act as corridors along which animals can move between regions of suitable habitat (e.g. for feeding). Corridors are also important for the establishment and expansion of some plant species.
   - Hedgerows provide habitat for the predators of prey species. This may also benefit the farmer by helping to keep pest species under control.

3. (a) and (b) any disadvantages of:
   - Hedgerows hamper effective use of some farm machinery.
   - Hedgerows take up space that could otherwise be used for grazing or crop production.
   - Hedgerows provide habitat for competitors to grazing livestock (e.g. hares) and predators (foxes).

4. (a) Habitat loss (hedgerows and woodland areas).
   (b) Decline in abundance and diversity of food sources associated with habitat loss.

5. (a)
   - Hedgerow legislation to preserve existing hedgerow habitats.
   - Policies to preserve or restore woodland cover in previously wooded areas (afforestation).
   - Schemes (with financial incentives) to encourage environmentally sensitive farming practices.

   (b) Biodiversity estimates would determine the biodiversity levels in certain habitats (e.g. with hedegrows or without). This would identify areas which required implementation of the biodiversity policies mentioned above, and also provide a baseline to measure the effectiveness of the strategies once implemented.

Biodiversity and Conservation (page 295)

1. **Vulnerable** species are those that are being reduced in numbers or range to such an extent that they are likely to become endangered in the near future. **Endangered** species are so reduced in numbers (populalltion size) that they are likely to become extinct unless immediate action is taken to arrest their decline. **Extinct** species are those with no living individuals left.

2. (a) and (b) any two of the the following:

   - Each species has a functional role in its ecosystem. The loss of a species may upset the dynamics, stability, and long term viability of an ecosystem.
   - Some endangered organisms may be valuable to humans as food sources, medicines, or sources of chemicals for which uses have yet to be identified.
   - From a humanitarian point of view, humans have no right to exterminate another species and should prevent extinction if they have the power to do so.

   Note: Some people argue that extinction is a natural process anyway and that it is a waste of money trying to save species that are doomed to become extinct. Other people say that we have a moral responsibility to act as custodians of life on the planet, as human activities have such a profound effect on them.

3. (a) CITES: Aims to ensure that international trade in species of wild plants and animals does not threaten their survival. Species are placed into CITES categories based on their ability to sustain trade (take). The categories range from a complete ban on trade to varying degrees of regulated take
   (b) Gene banks: Provide a store of genetic diversity from wild stocks so that the genetic diversity is preserved in the advent of species loss or decline. Using modern reproductive technologies, gene banks can be used to boost the genetic diversity of inbred populations of endangered species.
   (c) Habitat restoration: Aims to restore habitat to the state where it can support and maintain its previous flora and fauna. Suitably restored habitat may be used to enable expansion of populations threatened by loss of suitable habitat in specific areas.
   (d) Habitat protection: Aims to protect existing areas of high conservation value habitat. Habitat protection is an important part of endangered species programmes: it enables endangered species to be managed *in situ* and without disturbance and supports captive breeding programmes (captive bred species cannot successfully be returned to the wild if the habitat is degraded).
   (e) Captive breeding and release programmes: Aims to restore numbers of endangered species to levels where they can be released into areas of suitable habitat and (hopefully) survive and breed there. For very rare species (e.g. where there are fewer than 100 individuals left) captive breeding may offer the only chance to prevent rapid extinction.

4. ***In situ*** conservation measures engage a whole-ecosystem management approach to saving species within its own habitat. Methods include protecting, cleaning or restoration of the habitat, and protection of the endangered species from predators. If whole-

ecosystem restoration is successful it offers a good chance of species recovery, even for critically endangered species. It has the advantages of less disturbance to the species involved, it by-passes the need for captive breeding (which is unsuccessful for some species), and it offers a greater chance of long term success because habitat restoration goes hand in hand with species management.

Ex-situ conservation methods are often employed when species numbers become critically low or *in-situ* methods are not working. Ex-situ conservation involves removal of the endangered species from its natural habitat and moves it to a new location, often a protected or controlled area. This method rarely saves a species from extinction, and are often costly and labour intensive. Because the breeding stock is also limited, genetic diversity of the species may become compromised.

National Conservation (page 297)

1. (a) **National Parks**: Areas of substantial size protected and reserved for public enjoyment. Human activity (farming, quarrying etc.) is not excluded but maintained at a level appropriate to maintaining the desirable and characteristic features of the area.

   (b) **NNR**: Smaller and more numerous areas than National Parks, identified under government legislation specifically because of their importance in protecting biological diversity in Britain. In addition to their conservation role, they also function as areas for research and education. They are protected by law but are not guaranteed a protected future (this depends on individual landowners).

   (c) **SSSIs**: Designated by government agencies because of their particular scientific value in terms of their flora, fauna, geologic or physiographic features. Important in protecting natural areas because planning authorities, landowners, or occupiers must be informed of potentially damaging activities within the area. *NOTE: development within these areas does occur however (e.g. for roads or leisure/tourism) and their protection is not complete (esp. as funding for their protection is not assured).*

   (d) **ESAs**: Areas of national significance (determined by the Minister of Agriculture) whose conservation depends on adopting/maintaining/ extending a particular farming practice. Farmers are paid to manage the land in such a way as to conserve the desirable features, e.g. sheep grazing to maintain a chalk grassland, or reducing drainage to maintain water meadows. These managed habitats contain much of Britain's characteristic flora and fauna.

   (e) **The EU Habitats Directive**: A directive for the conservation of natural habitats and wild flora and fauna. Aims to ensure preservation or restoration of biodiversity in the European territory.

   (f) **Natura 2000**: An ecological network of special areas of high conservation value. These areas aim to conserve natural habitats and species of plants and animals that are rare, endangered, or vulnerable in the European Community. Includes Special Areas of Conservation (associated with plants and animals other than birds) and Special Protection Areas (associated with birds).

   Note: Access the web sites pertaining to these organisations from Biozone's **Bio Links** (select sites under **Conservation**).

2. (a) **Conservation of biodiversity**: RSPB, Woodland Trusts and The Wildlife Trusts purchase and manage nature reserves and conservation land for the purposes of maintaining high conservation value habitat (and as a corollary also biodiversity). RSPB is concerned primarily with bird habitat (but this benefits other species also) but has developed wider roles in lobbying for the protection of biodiversity and the promotion of tougher laws for environmental protection. The Woodland Trust is primarily concerned with the conservation of broad leaved woodlands and is also involved in habitat creation. The Wildlife Trusts manage their reserves on sustainable principles with the aim being to enrich wildlife generally.

   (b) **Protection of habitat**: Purchases of land enable land to be set aside for conservation purposes (i.e. not developed for other roles). Careful management helps maintain the integrity and natural features of the conservation estate.

   (c) **Environmental restoration**: The Woodland Trust aims not only to protect existing woodlands but to create new woods through replantings. The aim is to emulate natural processes as far as possible in the types of habitat creation schemes involved.

   (d) **Education and promotion of conservation aims**: NGOs have a vital role in raising public awareness of natural habitats, flora, and fauna, and their value. Education and increased public awareness are the primary ways in which to achieve long term public support for conservation aims. Activities are made known to the public and the public benefits from the activities of the NGOs (e.g. The Woodland Trust maintains facilities to encourage environmentally friendly activity in their reserve areas).

3. (a) The set-aside scheme offers farmers a financial incentive to take at least 20% of their land out of production and leave fields uncultivated.

   (b) Farmers receive compensation for the loss of income resulting from retiring land from production.

Measuring Diversity in Ecosystems (page 299)

1. (a) **Species richness** measures the number of species within an ecosystem, whereas **species evenness** describes how equally the species are distributed within an ecosystem.

   (b) Both measures are important when considering species conservation. Species richness could give an indication of ecosystem stability, and therefore how at-risk particular species may be. Species evenness provides an indication of the species distribution (a limited distribution or a distribution where individuals are widely separated may indicate the species is at risk).

2. Sampling must be carried out in an unbiased manner which provides a true representation of the ecosystem. This is usually achieved by random sampling techniques. The sample size must be large enough to gather a true representation of the species present, and the sampling method used must be suitable for capturing information of the species likely to be present.

3. High diversity systems have a greater number of biotic interactions operating to buffer them against change (the loss or decline of one component (species) is

less likely to affect the entire ecosystem). With a large number of species involved, ecosystem processes, such as nutrient recycling, are more efficient and less inclined to disruption.

4. **Keystone species** are pivotal to some important ecosystem function such as production of biomass or nutrient recycling. Because their role is disproportionately large, their removal has a similarly disproportionate effect on ecosystem function.

5. Species diversity index used in (any of):
   - Comparisons of similar ecosystems which have been subjected to (beneficial or detrimental) human influence (e.g. restoration or pollution).
   - Assessment of the same ecosystem before and after some event (fire, flood, pollution, environmental restoration).
   - Assessment of the same ecosystem along some environmental gradient (e.g. distance from a point source of pollution).
   - Assessment of the biodiversity value of an area for the purposes of management or preservation (tends to be a political lobbying point).

6. (a) DI = 37 x 36 ÷ ((7 x 6) + (10 x 9) + (11 x 10) + (2 x 1) + (4 x 3) + (3 x 2)) = 1332 ÷ 262 = 5.08
   (b) Without any frame of reference (e.g. for a known high or low diversity system), no reasonable comment can be made about the diversity of this ecosystem. Herein lies the problem with an index that has no theoretical upper boundary.

CITES and Conservation (page 301)

1. In 1989, the African elephant was placed in Appendix 1 of CITES, which imposed a ban on trade in living or dead material from elephants.

2. (a) A limited legal trade in ivory has resulted from a policy of management and quota operation. Removal of the ban on ivory has allowed the rural communities of these countries to earn money from the controlled exploitation of their wildlife. (Advocates of this claim that it has dramatically increased the amount of land given over to wildlife, as the returns from wildlife exploitation have exceeded those from cattle).
   (b) Any two of:
   - Quota systems can be abused (and have been in the past, with illegal hunting continuing).
   - As returns from ivory increase there will also be pressure to extend the quota above what individual elephant populations can sustain.
   - As ivory is traded, there will be pressure to illegally bring in ivory (for trade) poached from vulnerable populations outside quota countries.

The Modern Theory of Evolution (page 303)

Main points are:

Erasmus Darwin (1731 – 1802):
Charles Darwin's grandfather. English physician and scientist who proposed a theory of evolution by the *Inheritance of Acquired Characteristics*. Probably had a great influence on the thinking of Charles Darwin.

John Baptiste de Lamarck (1744 – 1829):
Contributed greatly to the classification of invertebrates. Best known for a mistranslation of his 1809 statement that species had evolved by adapting to a "need". This was translated into English as "desire" and he was ridiculed unjustly by English speaking scientists. (Much quoted example was that he was supposed to have suggested that giraffes could have stretched their necks by *wanting* to).

Thomas Malthus (1766 – 1834):
Wrote an important essay on controls on population growth that helped inspire the evolutionary theories of both Darwin and Wallace. Malthus proposed that the human population would be wiped out unless its birthrate was limited.

Herbert Spencer (1820 – 1903):
Proposed the concept of Survival of the Fittest that he first used in his book *Principles of Biology* (1864). This idea was adopted by Darwin and Wallace in their theory of *Evolution by Natural Selection*.

Charles Lyell (1797 – 1875):
British geologist famous for promoting the work of James Hutton of *Geological Uniformitarianism* (that the same agencies are at work in nature today, operating at the same intensities as they have always done throughout geological time e.g. erosion and sedimentation rates).

Alfred Russel Wallace (1823 – 1913):
Wallace jointly proposed the theory of evolution by natural selection with Darwin. One of the first people to map the distribution of living things leading him to propose the world is divided up into biogeographical zones. He wrote to Darwin of his ideas on evolution by natural selection, spurring Darwin on to publish *The Origin of Species*.

Gregor Mendel (1822 – 1884):
Explained the process of inheritance involving "particles" that are passed on undamaged through the sex cells to the next generation. This was a radical departure from the view of the day that inheritance involved a mere *blending* of parental characteristics.

August Weismann (1834 – 1914):
German biologist who is regarded as the father of modern genetics. He discredited the theory that acquired characteristics could be inherited. He was the first to propose that chromosomes are the basis of heredity.

Theodosius Dobzhansky (1900 – 1975):
A Russian biologist working in the area of population genetics. One of the architects of the Modern Synthetic Theory of evolution. Published *Genetics and the Origins of Species* in 1937; it was a publication that marked the beginning of a new understanding of evolutionary biology. Dobzhansky studied isolating mechanisms and showed how speciation can occur.

Ernst Mayr (1904 – 2005):
A German evolutionary biologist who collaborated with Dobzhansky, Julian Huxley, and George Gaylord Simpson to formulate the modern evolutionary synthesis. Worked on speciation in animals and defined different types of speciation mechanisms. Also proposed in the 1950s that rapid speciation events could occur. This became important for later ideas on punctuated equilibrium.

Julian Huxley (1887 – 1975):
English evolutionary biologist who worked on ritualisation behaviour, neoteny, and allometric growth before collaborating with Dobzhansky, Mayr, and George Gaylord Simpson to formulate the modern evolutionary synthesis. Wrote: Evolution: *The Modern Synthesis* in 1942 after which the new theory was named.

J.B.S. Haldane (1892 – 1964):
English geneticist who contributed to the Modern Synthetic Theory of evolution. Remembered most as an innovative

pioneer in population genetics, a field which reshaped modern evolutionary biology.

Sewall Wright (1889 – 1988):
American population theorist and one of the architects of the new evolutionary understanding. Together with Haldane and Fisher, Wright gave evolutionary biology a mathematical basis by working out the mathematical principles of population genetics. This transformed Darwinism into a 20th century science. Best known for his contributions to knowledge of evolution in small populations and the Founder effect (called the Sewall-Wright Effect).

R. A. Fisher (Sir) (1890 – 1962):
English statistician and geneticist who contributed to the Modern Synthetic Theory of evolution. He showed that Mendel's work and Darwin's ideas on natural selection were in agreement, not conflict as some had believed. Made major contributions also to the development of statistical ideas and to knowledge of human inheritance. Note that although Haldane, Fisher, and Wright made related contributions to the knowledge of population genetics, they were not collaborators and did not view themselves as such.

The Species Concept (page 304)

1. Behavioural (they show no interest in each other).
2. Physical barrier; sea separating Australia from SE Asia.
3. The red wolf is rare and may have difficulty finding another member of its species to mate with.
4. The populations on the two land masses, which have identical appearance and habitat requirements, were connected relatively recently by a land bridge during the last ice age (about 18 000 years ago). This would have permitted breeding between the populations. Individuals from the current populations have been brought together and are able to interbreed and produce fertile offspring.

Variation (page 305)

1. Continuous variation is characterised by an exceedingly large number of phenotypic variations (so that a large sample of the population would exhibit a normal distribution for the trait in question). Such traits are determined by a large number of genes and are also frequently influenced by environment, e.g. hand span, weight, skin colour. Discontinuous variation is characterised by a limited number of phenotypic variants. Such traits are determined by a single gene and include features such as chin dimple (present/absent) and blood groups (A, B, AB, O).
2. (a) Wool production: Continuous
   (b) Kernel colour: Continuous
   (c) Blood groups: Discontinuous
   (d) Albinism: Discontinuous
   (e) Body weight: Continuous
   (f) Flower colour: Discontinuous
3. Environmental influence expected on: wool production (a), kernel colour (b), and body weight (e).
4. Student's own plot. Shape of the distribution is dependent on the data collected. The plot should show a **statistically normal distribution** if sample is representative of the population and large enough.
   (a) Continuous distribution, normal distribution, or bell shaped curve are all acceptable answers if the data conform to this pattern.
   (b) Polygenic inheritance: Several (two or more) genes are involved in determining the phenotypic trait. Environment may also have an influence, especially if traits such as weight are chosen.
   NOTE: A large enough sample size (30+) provides sufficient data to indicate the distribution. The larger the sample size, the more closely one would expect the data plot to approximate the normal curve (assuming the sample was drawn from a population with a normal distribution for that attribute).
5. (a) **Plant A**: The observed phenotype (prostrate) of this species is not due to genetic factors, but to the effect of climate on growth patterns. In the absence of a harsh environment, the plant reverts to its normal growing habit.
   Plant B: The low growing phenotype of this species is controlled by genes (not environmental factors).
   (b) Plant A is most likely to show clinal variation.

Adaptations and Fitness (page 307)

1. Adaptive features are genetically determined traits that have a function to the organism in its environment. Physiological 'adaptation' (acclimatisation) refers to the changes made by an organism during its lifetime to environmental conditions (note that some adaptive features do involve changes in physiology).
2. Shorter extremities are associated with colder climates, whereas elongated extremities are associated with warmer climates. The differences are associated with heat conservation (shorter limbs/ears lose less heat to the environment).
3. Large body sizes conserve more heat and have more heat producing mass relative to the surface area over which heat is lost.
4. **Snow bunting** adaptations:
   (a) **Structural**: Large amount of white plumage reduces heat loss, white feathers are hollow and air filled (acting as good insulators).
   (b) **Physiological**: Lay one or two more eggs than (ecologically) equivalent species further south producing larger broods (improving breeding success), rapid moult to winter plumage is suited to the rapid seasonal changes of the Arctic.
   (c) **Behavioural**: Feeding activity continues almost uninterrupted during prolonged daylight hours (allowing large broods to be raised and improving survival and breeding success), migration to overwintering regions during Arctic winter (escapes harsh Arctic winter), will burrow into snow drifts for shelter (withstand short periods of very bad weather), males assist in brood rearing (improved breeding success).
5. Extra detail is (italics) provided as explanation:
   (a) Structural (*larger, stouter body conserves heat*).
   (b) Physiological (*concentrated urine conserves water*).
   (c) Behavioural (*move to favourable sites*).
   (d) Physiological (*higher photosynthetic rates and water conservation*).
   (e) Structural (*reduction in water loss*).

(f) Behavioural and physiological (*hibernation involves both a reduction in metabolic rate and the behaviour necessary to acquire more food before hibernation and to seek out an appropriate site*).
(g) Behavioural (*increase in body temperature*).

Darwin's Theory (page 309)

1. Natural selection can provide the means for species change over time because natural selection will always favour the most adaptive phenotypes (therefore genotypes) at the time. More favourable phenotypes will have greater reproductive success and will become proportionally more abundant in the population. Over time, favourable phenotypes will predominate and the unfavourable phenotypes will become very rare.

Natural Selection (page 310)

1. (a) **Directional** selection acts against phenotypes at one extreme of the phenotypic range (such that the phenotypic mean shifts in the other direction) and is associated with gradual environmental change. **Disruptive** selection, in contrast, is associated with fluctuating environments and favours phenotypic variants at each extreme of the phenotypic range, such that intermediate phenotypes are eliminated.
   (b) Directional selection is most often associated with evolution (change in the gene pool over time). Disruptive selection may lead to evolution if one or both of the phenotypic extremes are then subjected to new (different) selection pressures.
2. A shift in environmental conditions can favour one particular phenotype (or phenotypes) at the extreme of the phenotypic range. The phenotypes favoured in the new conditions will have greater reproductive success. Natural selection will act against the reproduction of the unfavourable phenotypes in the new conditions.
3. In a population of variably coloured snails, the environmental conditions could change so that pale forms became more susceptible to predation. The darker forms, being better camouflaged, would have better survival and reproduction. Natural selection would favour these darker phenotypes and act against the pale forms, which would have poor survival and reproduction. Over time, the dark forms would predominate and lighter forms would become very rare.

Selection For Human Birth Weight (page 311)

Sample data and graph below. Note: For the construction of weight classes, it is necessary to have a range of weight categories that do not overlap. The data collected should be sorted into weight classes of: 0.0-0.49, 0.50-0.99, 1.0-1.49, 1.5-1.99, etc.

SAMPLE DATA*: Use these data if students are unable to collect from local sources*

3.740	3.830	3.530	3.095	3.630
1.560	3.910	4.180	3.570	2.660
3.150	3.400	3.380	2.660	3.375
3.840	3.630	3.810	2.640	3.955
2.980	3.350	3.780	3.260	4.510
3.800	4.170	4.400	3.770	3.400
3.825	3.130	3.400	3.260	4.100
3.220	3.135	3.090	3.830	3.970
3.840	4.710	4.050	4.560	3.350

3.380	3.690	1.495	3.260	3.430
3.510	3.230	3.570	3.620	3.260
3.315	3.230	3.790	2.620	3.030
3.350	3.970	3.915	2.040	4.050
3.105	3.790	3.060	2.770	3.400
1.950	3.800	2.390	2.860	4.110
1.970	3.800	4.490	2.640	3.550
4.050	4.220	2.860	4.060	3.740
4.082	3.000	3.230	2.800	4.050
4.300	3.030	3.160	3.300	2.350
3.970	2.980	3.550	3.070	2.715

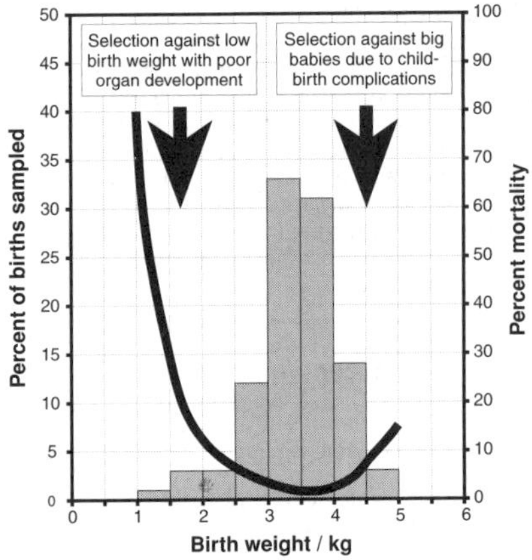

1. Normal distribution (bell-shaped curve), probably with a skew to the left.
2. 3.5 kg (taken from the table: only 2% mortality).
3. Good correlation. Lowest frequencies of surviving birth weights correspond to birth weights of highest mortality.
4. Selection pressures operate at extremes of the range: premature babies have reduced survival capacity because their body systems are not fully developed; large babies present problems with delivery as the birth canal can only cope with babies up to a certain size. **Note**: Mothers with diabetes often have very large babies and before adequate medical intervention, this often led to the death of both mother and baby.
5. Medical intervention has improved the survival rates of very premature babies (as small as 1.5 kg), but this has not historically been the case. Caesarean deliveries have also allowed larger babies to be born. Note that this technology is available to wealthy societies thereby reducing the effect of this selection pressure. Developing countries still experience this selection pressure.

Stages in Species Development (page 312)

1. Gene flow becomes reduced and the two populations evolve in different directions (until gene flow ceases).
2. Significant geographical barriers would have prevented much gene flow. The large distances between the sparsely populated communities would have made contact a rare event, even for nomadic people. It may have been seen as desirable to introduce 'new blood' into a group when meeting others and women may

have been exchanged. More 'hostile gene flow' may have occurred if sexual intercourse was forced upon women during attacks between warring peoples.

3. (a) Geographical barrier in the form of vast oceans.
   (b) Some ducks may still be deterred by mating behaviour that does not match exactly that required by their own species.

Fossil Formation (page 313)

1. (a) Pyritisation: Iron pyrite replaces hard remains of the dead organisms.
   (b) Amber: Conifer resin or gum traps insects or other small invertebrates and then hardens.
   (c) Petrification: Wood is silicified: silica from weathered volcanic ash is incorporated into the decayed wood.
   (d) Phosphatisation: Bones and teeth are preserved in phosphate deposits.
   (e) Tar pit: Organisms fall into a tar pit (mix of sand and tar) and are trapped there. Their remains become embedded in the matrix of sand and tar.

2. Decay.

3. **Transitional fossils** are those processing a mixture of the characteristics of two different, but related, taxonomic groups. They are important because they indicate that one group may have given rise to the other by evolutionary processes.

The Fossil Record (page 315)

1. (a) Any of: horse, elephant, pig, numerous dinosaur groups, trilobites.
   (b) Any of: tuatara, coelacanth, gingkoes (ancient conifers).

2. Fossilised plants and animals from the fossil record can be compared to living species today thus giving an insight into how they have evolved over time.

3. (a) Layer A (b) Layer H (c) Layer I (d) Layer O

4. (a) Layer E
   (b) It has the same relative position in the sequence of layers and it contains fossils typical of the layer.

5. These rocks may be so old that large organisms with hard body parts were not present to be fossilised (i.e. they may have had soft body parts that decomposed).

6. (a) Layers C and F (b) Layer J

7. Any three of:
   Radiometric dating: Measuring radio-isotope ratios.
   Fossil correlation: Matching up fossil community types with those at another location with a known date.
   Palaeomagnetism: Measuring the magnetic alignment of the rock and correlating this with known magnetic pole directs in the earth's past.
   Fission track analysis: Measuring the number of tracks caused by particles in rock crystals.

8. (a) Layer A: 0 – 80 million years old (< 80 mya)
   (b) Layer C: 80 – 270 million years old
   (c) Layer E: 270 – 375 million years old
   (d) Layer G: Older than 375 million years old
   (e) Layer L: 270 million years old
   (f) Layer O: Older than 375 million years old

Dating a Fossil Site (page 317)

1. **Occupation horizons** (layers in the soil profile with evidence of human occupation) are indicators of human activity some time in the past. Each one is the ancient living floor of the site and can provide information about the type of lifestyle and cultural development of the human occupants.

2. (a) Older than 18 500 but younger than 45 000 years.
   (b) The upper surface is about 18 500 years since it has the hearth (which has been dated) near the top.

3. (a) Pottery bowl: Thermoluminescence.
   (b) Skull: Radiocarbon-14, Uranium-thorium, Electron spin resonance.
   (c) Hearth: Radiocarbon-14, Thermoluminescence.
   (d) Tooth: Electron spin resonance.

4. Palaeoanthropologists document **all** of the remains at a fossil site to maximise the recovery of information. This allows for cross referencing between finds, i.e of hominin and non-hominin species, and gives a more accurate picture of the past environment and ecology.

5. Palaeoanthropology is a multidisciplinary science encompassing palaeontology, geology, prehistoric archaeology, molecular biology, and behavioural science amongst others. Different scientific disciplines provide information on quite different aspects of a site's physical and biological environment and allow any site of hominin activity to be examined in a holistic way. Working in this way, scientists can piece together a more accurate interpretation of past ecology and behaviour.

Darwin's Finches (page 319)

1. Main factors contributing to adaptive radiation: absence of competitors on the Galapagos and, partly as a consequence of this, a wide diversity of niches available for exploitation. **Note**: Radiation such as this would also have required a relatively unspecialised founding species with a certain amount of genetic plasticity.

DNA Hybridisation (page 320)

1. The similarity of DNA from different species can be established in a rudimentary way by measuring how closely single strands from each species mesh together. The more similar the DNA, the harder it is to separate them.

2. (a) Chimpanzee (b) Galago

3. (a) 7 – 8 (b) Approx. 12

4. 45 million years ago.

Immunological Studies (page 321)

1. **Immunological studies** have been used as a crude means of determining the similarity of proteins between species. The technique uses the ability of the immune system of a mammal to recognise foreign proteins. Results have confirmed schemes of evolutionary relationships based on anatomical evidence.

2. (a) 60 (b) 25 – 30

3. Distantly related (branched off 35-40 million years ago).

4. 25 million years ago.

Other Evidence for Evolution (page 322)

1. **Amino acid sequences**: Chimpanzees have identical amino acid sequences to humans for some proteins, while gorillas vary only slightly. Other primates, such as monkeys, have many more differences in the chemical makeup of their proteins.

2. **Comparative embryology**: Animals that are thought to share closer evolutionary relationships are found to have an embryonic development that is similar to a later stage.

3. **DNA profiling** and **DNA sequencing** are two methods that have lead to a revolution in evolutionary biology. Application is based on the assumption that more closely related organisms will have more similar DNA. The longer two species have been separated in time, the greater the opportunity for change to the DNA through mutation. DNA techniques have also been used to determine the similarity of mitochondrial DNA in different racial groups in humans and gave rise to the 'Eve hypothesis' for the origin of modern humans.

Homologous Structures (page 323)

1. (b) Human arm: Modified for tree climbing (flexible joints) and improved dexterity of fingers.
   (c) Seal flipper: Modified to increase surface area and streamlined to function as a paddle.
   (d) Dog foot: Modified for swift running in pursuit of prey. Walks on toes, long limbs to provide lengthened, running stride.
   (e) Mole forelimb: Short and strong limb. Shovel-like paw with sharp claws for digging and propelling itself underground.
   (f) Bat wing: Modified into a wing for flying. Very long metacarpals and fingers stretch the skin into a wing.

2. The limbs all share the same basic bone anatomy, although highly modified in some cases. It is possible to match bone for bone, but also recognise how individual bones or bone groups have changed to better perform a new function for the animal.

3. Innate or genetically determined behaviour among animals is inherited in the same way as structural features. While some behaviour is learned, this tends to occur within, rather than between species.

Vestigial Organs (page 324)

1. When an organism adopts a new niche, exploits a new habitat, or takes on 'new' way of doing something, some existing structure may become redundant. Rather than helping the organism to exploit its new way, they may hinder it. Selection pressures may then act against the organs being large. Even if there is no direct selection pressure, the organ may still regress with time as less energy is put into growing a little used body part.

2. The genes that code for it are still present and will continue to express themselves to produce the structure. What is required (for its loss) is an accumulation of mutations that will cause the switching off of the relevant genes altogether.

3. It is possible to see a gradual reduction in the size of the vestigial organ as it progresses from the early ancestor, through **transitional forms**, to modern forms.

Antibiotic Resistance (page 325)

1. (a) Antibiotic resistance arises in a bacterial population as result of mutation. Some bacteria can also acquire these changes in DNA (conferring resistance) by transfer of genes between bacteria by conjugation (horizontal evolution).
   (b) Resistance can become widespread as a result of (1) transfer of genetic material between bacterial (horizontal evolution) or by (2) increasing resistance with each generation a result of natural selection processes (vertical evolution). In the latter case, the antibiotic provides the environment in which selection for resistance can take place. It is exacerbated by overuse and misuse of antibiotics.

2. Widespread antibiotic resistance has implications for the treatment and control of what have been, in the past, quite easily treated diseases. Tuberculosis is one good example. Historically, it was effectively treated with antibiotics, but complacency over its control has lead to increasing multiple drug resistance in the *Mtb* population and a resurgence in the number of TB cases. This has huge implications for public health because more people live with (resistant forms of) the disease and spread it to more people as a result. In addition, the costs associated with treating TB are now also much higher.

   Another case is the increasing incidence of hospital-associated infections involving resistant strains of common bacteria such as *S.aureus*. Resistant infections acquired during hospital stays in immune-compromised patients prolong recovery, increase the number of deaths related to incidental infections, and greatly increase the costs of treating patients.

   In general, increasing resistance increases the costs lowers the efficacy of treating disease.

Insecticide Resistance (page 326)

1. (a) Generation times in most pest insects are very short, so successive generations can be exposed (and respond) to the selection pressure of the insecticide in a short space of time.
   (b) Increasingly higher dose rates of insecticide are used to combat increasing resistance (= lack of effective control). This drives the selection process and increases the rate at which resistance spreads in the population.

2. After each insecticide application, only the most resistant individuals remain to reproduce. These offspring are, in turn, exposed to the insecticide. Again, the most resistant survive. With repeated applications of insecticide, the proportion of resistant individuals in the population increases.

3. Resistance to insecticides has implications for how humans effectively control the incidence and spread of vector-borne infectious diseases. Such diseases can potentially be effectively controlled, at low cost, over wide areas by controlling the populations of insect vectors (in the case of malaria, this is various mosquito

species). Indiscriminate use of insecticides (e.g. DDT) to control agricultural pests leads to widespread resistance in insect populations generally and reduces (or eliminates) the efficacy of the insecticide in the control of disease. This, in turn, leads to increasing incidence of the vector and the disease, higher disease control costs, and higher costs in treating the disease in infected populations (because incidence is higher).

Components of an Ecosystem (page 328)

1. A **community** is a naturally occurring group of organisms living together as an ecological entity. The community is the biological part of the ecosystem. The **ecosystem** includes all of the organisms (the community) and their physical surroundings.

2. The **biotic factors** are the influences that result from the activities of organisms in the community whereas the **abiotic** (physical) **factors** are the influences of the non-living part of the community, e.g. climate.

3. (a) Population (b) Ecosystem (c) Community (d) Physical factor

Habitat (page 329)

1. An organism will occupy habitat according to its range of tolerance for a particular suite of conditions (temperature, vegetation and cover, pH, conductivity). Organisms will tend to occupy those regions where all or most of their requirements are met and will avoid those regions where they are not. Sometimes, a single factor, e.g. pH for an aquatic organisms, will limit occupation of an otherwise suitable habitat.

2. (a) Most of a species population is found in the optimum range because this is the zone where conditions for that species are best; most of the population will select that zone.
   (b) The greatest constraint on an organism's growth within its optimum range would be intraspecific competition (or competition with different species with similar niche requirements).

3. In a marginal niche, any of the following might apply:
   - Physicochemical conditions (e.g. temperature, current speed, pH, conductivity) might be sub-optimal and create stress (therefore greater vulnerability to disease).
   - Food might be more scarce or of lower quality.
   - Mates might be harder to find.
   - The area might be more exposed to predators.
   - Resting, sleeping, or nesting places might be harder to find and/or less suitable in terms of shelter or safety.
   - Competition from other better-adapted species might be more intense.

Ecological Niche (page 330)

1. (a) The realised niche constitutes only a small proportion of the niche range which an organism can occupy. It is mainly determined by interspecific competition, so can expand or contract depending upon the level of competition at any one time.
   (b) Competition with other species may prevent the organism from exploiting all resources it is adapted to use. Competition forces species to occupy a **realised niche** that is narrower than their fundamental niche.

2. An organism will occupy habitat according to its range. Intraspecific competition tends to broaden niches as scramble competition forces individuals to move outside their optimum resource range. Interspecific competition tends to make niches narrower because it encourages niche differentiation and resource specialisation (compartmentation of the available resources) to minimise scramble competition.

Food Chains and Webs (page 331)

1. (a) Each successive trophic level has less energy.
   (b) Energy is lost by respiration as it is passed from one trophic level to the next.

2. A food chain comprises a sequence of organisms, each of which is a source of food for the next. Food chains in ecosystems are organised according to trophic levels; the feeding levels that energy passes through as it proceeds through the food chain. Organisms are assigned a category according to the trophic level they occupy. Producers form the first trophic level, 1st order consumers (primary consumers) eat producers (i.e they are herbivores), 2nd order consumers (secondary consumers) eat herbivores (i.e. they are carnivores) etc. Organisms may occupy more than one trophic level depending on their diet. Detritivores and decomposers obtain energy from all other trophic levels and are therefore not assigned a trophic level.

3. (a)-(e) Some food chain examples as below:
   - Macrophyte → carp → pike
   - Algae → zooplankton → diving beetle
   - Algae → zooplankton → stickleback → pike
   - Macrophyte → great pond snail → herbivorous water beetle → stickleback → pike
   - Algae → mosquito larva → *Hydra* → dragonfly larva → carp → pike
   - Macrophyte → herbivorous water beetle → carp → pike
   - Algae → zooplankton → *Asplanchna* → leech → dragonfly larva → carp → pike

4. Food web solution: See the top of the next page.

   Trophic levels are indicated by the letter T and the number(s) of the level(s) occupied. Note: the trophic level a species occupies will depend on the trophic position of its food items. For example, the carp occupies several different trophic levels, since it feeds on macrophytes, and on both primary and tertiary consumers. The tertiary consumers that the carp eats will also be feeding at a number of different levels, hence the complexity of food webs and the difficulty in accurately representing them.

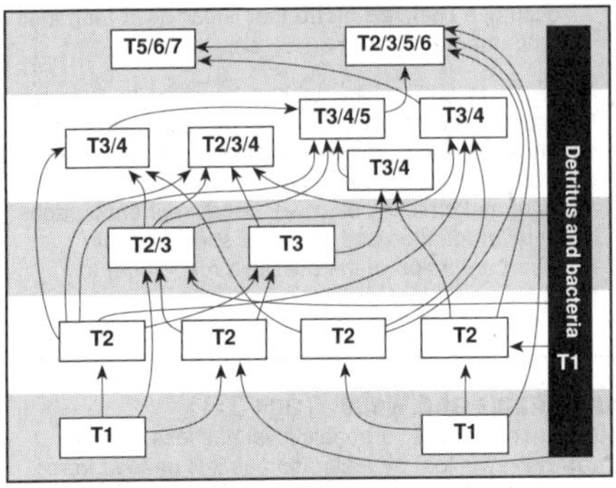

Energy Inputs and Outputs (page 333)

1. **Producers** convert energy received from an inorganic source (usually sunlight) into a form that is accessible to consumer levels. **Consumers** depend on the energy stored in the chemical bonds of biological molecules (the fats, proteins, and carbohydrates of plant and animal tissues). They too transfer energy to other levels, but energy is lost with each transfer.

2. In a grazing food web, energy moves from producers (plants) to primary consumers (herbivores) and then to secondary consumers (carnivores). This chain of energy transfer can continue several times, but eventually ends. All these consumer groups provide energy to decomposer levels. In a detrital food web, producers provide energy as dead plant material, and the primary consumers are decomposer microbes such as bacteria and fungi. Energy flows back and forth between decomposers and detritivores but herbivores and carnivores do not feature.

3. Detritivores consume (ingest) detritus and, in doing so, speed up decomposition by increasing the surface area available to decomposer bacteria. Decomposers (bacteria and fungi) also use detritus as an energy source but digestion is extracellular (enzymes are secreted in fungi or bound to the cell surface in bacteria). These enzymes break down the detritus into constituent molecules for absorption so the breakdown is more complete than is the case with detritivores.

Energy Flow in an Ecosystem (page 334)

1. (a) 14 000 (c) 35
   (b) 180 (d) 100

2. Solar energy

3. A. Photosynthesis
   B. Eating/ feeding/ingestion
   C. Respiration
   D. Export (lost from the ecosystem to another)
   E. Decomposers and detritivores feeding on other decomposers and detritivores
   F. Radiation of heat to the atmosphere
   G. Excretion/egestion/death

4. (a) 1 700 000 ÷ 7 000 000 x 100 = 24.28%
   (b) It is reflected. Plants appear green because those wavelengths are not absorbed. Reflected light falls on other objects as well as back into space.

5. (a) 87 400 ÷ 1 700 000 x 100 = 5.14%
   (b) 1 700 000 - 87 400 = 1 612 600 (94.86%)
   (c) Most of the energy absorbed by the producers is **not** used in photosynthesis. This excess energy which is not fixed is lost as heat (although the heat loss component **before** the producer level is not usually shown on energy flow diagrams). **Note**: Some of the light energy that is absorbed through accessory pigments such as carotenoids widens the spectrum that can drive photosynthesis. However, much of accessory pigment activity is associated with photoprotection; they absorb and dissipate excess light energy that could damage chlorophyll.

6. (a) 78 835 kJ
   (b) 78 835 ÷ 1 700 000 x 100 = 4.64%

7. (a) Decomposers and detritivores
   (b) Transport by wind or water to another ecosystem (e.g. blown or carried in air/stream/river/ocean currents).

8. (a) Energy remains locked up in the detrital material and is not released.
   (b) Geological reservoir:

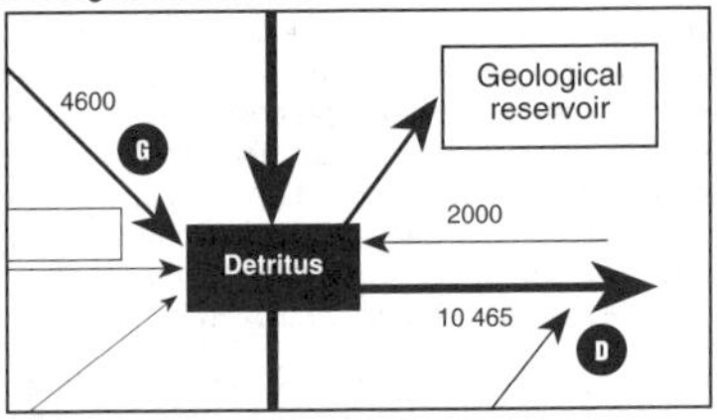

   (c) Oil (**petroleum**) and **natural gas**, formed from the buried remains of marine plankton. **Coal** and **peat** are both of plant origin; peat is partly decomposed, and coal is fossilized.

9. (a) 87 400 → 14 000: 14 000 ÷ 87 400 x 100 = 16%
   (b) 14000 → 1600: 1600 ÷14 000 x 100 = 11.4%
   (c) 1600 → 90: 90 ÷ 1600 x 100 = 5.6%

The Nitrogen Cycle (page 336)

1. (a)-(e) any of:
   - Decomposition or decay of dead organisms, to ammonia by decomposer bacteria (ammonification).
   - Nitrification of ammonium ions to nitrite by nitrifying bacteria such as *Nitrosomonas* ($NH_4^+ \rightarrow NO_2^-$)
   - Nitrification of nitrite to nitrate by nitrifying bacteria such as *Nitrobacter* ($NO_2^- \rightarrow NO_3^-$)
   - Denitrification of nitrate to nitrogen gas by anaerobic denitrifying bacteria such as *Pseudomonas* ($NO_3^- \rightarrow N_{2(g)}$)
   - Fixation of atmospheric nitrogen to nitrate by nitrogen fixing bacteria such as *Azotobacter* and *Rhizobium* ($N_2 \rightarrow NO_3^-$)
   - Fixation of atmospheric nitrogen to ammonia by nitrogen fixing cyanobacteria ($N_2 \rightarrow NH_3$)

SURA'S

- Biography Series -

MODERN INVENTORS

By

Archana Srinivasan

SURA BOOKS (Pvt) LTD.

Chennai • Bangalore • Kolkata • Ernakulam

Price: Rs.40.00

MODERN INVENTORS

By
Archana Srinivasan

First Edition : March, 2005
Second Edition : May, 2006
Size : ⅛ Demy
Pages : 96

Price: Rs.40.00

ISBN: 81-7478-636-8

SURA BOOKS (PVT) LTD.

Head Office:
1620, 'J' Block,
16th Main Road,
Anna Nagar,
Chennai - 600040.
Phones: 26162173, 26161099.

Branch:
XXXII/2328, New Kalavath Road,
Opp. to BSNL, Near Chennoth Glass,
Palarivattom,
Ernakulam - 682025.
Phone: 0484-3205797

Printed at T. Krishna Press, Chennai - 600 102 and
Published by V.V.K.Subburaj for Sura Books (Pvt) Ltd.,
1620, 'J' Block, 16th Main Road, Anna Nagar, Chennai - 600 040.
Phones: 26162173, 26161099. Fax: (91) 44-26162173.
email: surabooks@eth.net; website: www.surabooks.com

05 06 2000

Contents

BIOGRAPHY SERIES

MODERN INVENTORS

ISAAC SINGER

Singer was born in 1811 in Utica, New York, the son of Adam Singer, a Saxon immigrant to America, and his first wife Ruth. He entered a machinist's shop as an apprentice at the age of nineteen, but stayed there only a few months, leaving to become one of a touring group of actors. His income came alternately from work as a mechanic and as an actor. In 1830, he married Catherine Maria Haley.

In 1835, he moved with Catherine and their son William to New York City, working in a press shop. In 1836, he left the city as an advance agent for a company of players, touring through Baltimore, where he met Mary Ann Sponsler, to whom he proposed marriage. He returned to New York, where Catherine conceived and gave birth to a daughter, Lillian, in 1837.

After Mary Ann arrived in New York and discovered that Singer was already married, she and Singer returned to

Baltimore, presenting themselves as married. Their son Isaac was born in 1837.

First inventions

In 1839 Singer obtained his first patent for a machine to drill rock, selling it for $2,000. This was more money than he had ever had before, and in the face of financial success, he opted to return to his career as an actor. He went on tour, forming a troupe known as the "Merritt Players", and appearing onstage under the name "Isaac Merritt", with Mary Ann who also appeared onstage, calling herself "Mrs. Merritt". The tour lasted for about five years.

In 1844 Isaac took a job in a print shop in Fredericksburg, Ohio, but moved quickly on to Pittsburgh in 1846 to set up a woodshop for making wood type and signage. Here he developed and patented a "machine for carving wood and metal" on April 10, 1849.

At thirty-eight years old, with two wives and eight children, he packed up his family and moved back to New York City, hoping to market his machine there. He obtained an advance to build a working prototype, and obtained an offer to set up one of his machines in Boston. Singer went to Boston in 1850 to set the machine up at the shop of Orson C. Phelps, where Lerow and Blodgett sewing machines were being constructed. Orders for Singer's machine were not, however, forthcoming. Phelps asked Singer to look at the sewing machines, which were difficult to use and difficult to produce. Singer noted that the sewing machine would be more reliable if the shuttle moved in a straight line rather than a circle, with a straight rather than a curved needle.

Singer obtained financing, again, from George B. Zieber, becoming partners, with Phelps and him, in the "Jenny Lind Sewing Machine", named for Jenny Lind. Singer's prototype sewing machine became the first to work in a practical way. He received a patent in relation to improvements on the sewing machine on August 12, 1851. When eventually marketed, the machine was no longer the "Jenny Lind" but the Singer sewing machine.

Sewing machine design

Singer didn't invent the sewing machine, and never claimed to have done so. By 1850, when Singer saw his first sewing machine, it had been "invented" four times. All sewing machines before Walter Hunt's produced, a "chain stitch", had the disadvantage of easily unravelling. Hunt's machine produced a "lock stitch", as did all subsequent machines, including Lerow and Blodgett's, which Singer improved in Phelps's shop. Elias Howe independently developed a sewing machine and obtained a patent on September 10, 1846.

War broke out between Howe and Singer, with each claiming patent primacy. Singer set out to discover that Howe's improvements had been reinventions of existing technology, and found one of Hunt's old machines, which indeed created a lock-stitch with a shuttle. Hunt applied in 1853 for a patent, claiming priority to Howe's patent, issued some seven years earlier. A lawsuit, *Hunt v. Howe*, came to trial in 1854, and was resolved in Howe's favour. Howe then brought suit to stop Singer from selling Singer machines, and protracted litigation ensued.

I. M. Singer & Co.

In 1856, manufacturers Grover, Baker, Singer, Wheeler, and Wilson, all accusing the others of patent infringement, met in Albany, New York to pursue their suits. Orlando B. Potter, a lawyer and president of the Grover and Baker Company, proposed that, rather than suing their profits out of existence, they could pool their patents. This was the first patent pool, a process which enables production of complicated machines without legal battles over patent rights. They agreed to form the Sewing Machine Combination, but for this to be of any use they had to secure the cooperation of Elias Howe, who still held certain vital uncontested patents which meant he received a royalty on every sewing machine manufactured by any company. Terms were arranged, and Howe joined on. Sewing machines began to be mass produced: I. M. Singer & Co. manufactured 2,564 machines in 1856, and 13,000 in 1860 at a new shop on Mott Street in New York.

Sewing machines had until now been industrial machines, made for tailors, but smaller machines began to be marketed for home use. I. M. Singer expanded into the European market, establishing a factory in Clydebank, near Glasgow, controlled by the parent company, becoming one of the first American-based multinational corporations, with agencies in Paris and Rio de Janeiro.

Financial success

The financial success gave Singer the ability to buy a mansion on Fifth Avenue, into which he moved his second family. In 1860, he divorced his first wife, on the basis of her adultery with Stephen Kent. He continued to live with Mary Ann, until she spotted him driving down Fifth Avenue seated beside one Mary McGonigal, an employee, about whom Mary Ann had well-founded suspicions, for by this time Mary McGonigal had borne Isaac Singer five children. The surname Matthews was used for this family. Mary Ann (still calling herself Mrs. I. M. Singer) had her husband arrested for domestic violence. Singer was let out on bond and, disgraced, fled for London, taking Mary McGonigal with him. In the aftermath, another of Isaac's families was discovered: he had a "wife" Mary Eastwood Walters and daughter Alice Eastwood in Lower Manhattan, who both adopted the surname "Merritt". By 1860, Isaac had fathered and recognized eighteen children (sixteen of them remaining alive), by four women.

With Isaac in London, Mary Ann began setting about securing a financial claim to his assets by filing documents detailing his infidelities, claiming that though she had never been formally married to Isaac, that they were in fact wed under Common Law (by living together for seven months after Isaac had been divorced from his first wife Catherine). Eventually a settlement was made, but no divorce was granted. However, she asserted that she was free to marry and married John E.

Foster in Boston in 1862. Isaac now contended that in fact they had indeed been married under Common Law and accused Mary Ann of bigamy, and forced her to sign a renunciation of their prior financial settlement.

Singer then began seeing Mrs. Isabella Eugenie Boyer Summerville, said to have been a model for Bartholdi's Statue of Liberty, who left her husband and married Isaac on June 13, 1863.

Final years in Europe

In 1863, I. M. Singer & Co. was dissolved by mutual consent, with the business continued by "The Singer Manufacturing Company", enabling the reorganization of financial and management responsibilities. Singer no longer actively participated in the firm's day-to-day management, but served as a member of the Board of Trustees, and was a major stockholder.

He now began to increase his new family: he would eventually have six with his wife Isabella. Unable, probably because of Isaac's chequered marital past, to enter New York society, the family emigrated to Paris, never to return to the United States. Fleeing the Franco-Prussian War, they resided first in London, then in Paignton, (near Torquay) on the Devon coast where he built a large house, Oldway Mansion. He brought some of his other children to live there. Nine days after the wedding of his daughter Alice Merritt to William Alonso Paul La Grove, Isaac Singer died of an affection of the heart and inflammation of the wind-pipe in 1875. He was interred in Torquay cemetery.

Estate and legacy; his family after his death

Singer left an estate of about $14,000,000, and two wills disposing this between his family members, leaving some out for various reasons. Suits followed, with Mary Anne claiming to be the legitimate "Mrs. Singer". In the end Isabella was declared the legal widow. Isabella subsequently married a Belgian musician, Victor Reubsaet, who inherited the title Vicomte d'Estemburgh, and the Vatican title of Duke of Camposelice.

Isaac's 18th child Winnaretta Singer married Prince Louis de Scey-Montbeliard in 1887, when she was 22. After annulment of this marriage in 1891, she married to Prince Edmond de Polignac in 1893. She would become a prominent patron of french avant-garde music, e.g. Erik Satie composed his Socrate as one of her commissions (1918). As a lesbian she became involved with Violet Trefusis from 1923 on. Another of Isaac's daughters, Isabelle-Blanche (born 1869) married Elie, duc Decazes (Daisy Fellowes was their daughter). Isabelle committed suicide in 1896. A brother to Winnaretta and Isabelle, Paris Singer, had a child by Isadora Duncan. Another brother, Washington Singer, became a substantial donor to the University College of the South West of England, which later became the University of Exeter; one of the University's buildings is named in his honour.

LOUIS PASTEUR

We need no reminder that the foundations of our knowledge of health and disease were constructed by scientific giants who worked decades, even centuries, ago. It is with tributes such as the one today to Louis Pasteur that we pay homage to these great minds — to acknowledge their achievements and our indebtedness to them which we can never repay.

With certainty, one hallmark of Pasteur's research was not only the importance of his individual discoveries, but the overwhelming breadth of his accomplishment. Pasteur's long time collaborator, Emile Duclaux, wrote, "A mind.... of a scientific man is a bird on the wing; we see it only when it alights or when it takes flight.... We may by watching closely keep it in view, and point out just where it touches the earth. But why does it alight here and not there? Why has it taken this direction and not that in its flight toward new discoveries?"

Pasteur, himself, provided us with an answer: He believed that his research was "enchained" to an inescapable, forward moving logic. As we review today Pasteur's scientific discoveries we shall see the truth of this statement: how one discovery, one concept, led almost "inescapably" to another.

Pasteur's early life

Pasteur was born in Ole and grew up in the nearby town of Arbois, the only son of a poorly educated tanner, Jean Pasteur. Louis was not an outstanding student during his years of elementary education, preferring fishing and drawing to other subjects. In fact, young Louis' drawings suggested that he could easily have become a superior portrait Artist. His later drawings of friends done at college were so professional that Pasteur was listed in at least two compendia of 81 artists.

The Senior Pasteur, however, did not see his son ending up as an Artist, and Louis, himself, was showing increasing interest in chemistry and other scientific subjects. The highest wish Father Pasteur had for his son was that he complete his education in the local schools and become a professor in the college at Arbois. However the headmaster of the college recognized that Louis could do much better and convinced father and son that Louis should try for the Ecole Normale Suprieure in Paris. This most prestigious French University was founded specifically to train outstanding students for University careers in science and letters. And it was here that Pasteur entered and began his long journey of scientific discovery.

Crystallography

It may surprise some to learn that Pasteur, the father of microbiology and immunology, was a chemist who launched his memorable scientific career by studying the shapes of organic crystals. Pasteur was 26 years old, working for his doctorate in chemistry in the laboratory of Antoine Balard. Crystallography was just emerging as a branch of chemistry. His project was to crystallize a number of different compounds. Happily he started working with tartaric acid. Crystals of this organic acid are present in large amounts in the sediments of fermenting wine. Often one also found in the sediments in the wine barrels crystals of a second acid called paratartaric acid or "racemic acid". A few years earlier, the chemical compositions of these two acids, tartaric and paratartaric, had been determined. They were identical. But in solution there was a striking difference. Whereas tartaric acid rotated a beam of polarized light passing through it to the right, paratartaric acid did not rotate the light. This puzzled the young Pasteur. How could this be?

Pasteur refused to accept the notion that two compounds that had the same chemical composition yet acted so differently in respect to rotation of light could be identical. He was convinced that the internal structure of the two compounds must be different and this difference would show itself in the crystal form. The experts in this field had looked examined tartrate and paratartrate crystals but never saw a difference, perhaps because, as Duclaux thought, they believed that no difference could exist. Pasteur believed that there were differences and indeed found them!

Upon intense examination beneath his microscope, he saw that every crystal of pure tartaric acid looked like every other one. When he examined the paratartrate crystals, on the other hand, he saw two types of crystals, nearly identical but not quite! One type was the mirror image the other — the way the right hand mirrors the left hand. This was the difference he was looking for!

Pasteur then performed one of the simplest and yet most elegant experiments in the annals of chemistry. With a dissecting needle and his microscope, he separated the left and right crystal shapes from each other to form two piles of crystals. He then showed that in solution one form rotated light to the left, the other to the right. This simple experiment proved that the organic molecules with the same chemical composition can exist in space in unique stereospecific forms. And with this work did Pasteur launch the new science of stereochemistry.

To Pasteur this discovery had a deeper meaning. He proposed that asymmetrical molecules were indicative of living processes. In the broadest sense, he was correct. We know today that all of the proteins of higher animals are made up of only those amino acids that exist in the left-hand form. The mirror image right-hand amino acids are not used by human or animal cells. Likewise, our cells burn only the right-handed form of sugar, not the left-handed form that can be made in the test tube. It was the discovery of asymmetry of organic molecules that provided Pasteur with the "inescapable forward moving logic" that enchained him as he began his studies on alcoholic fermentation.

Alcoholic Fermentation

Pasteur served on the faculty of science of Dijon for a brief period and then was transferred to Strasbourg University where he continued his studies on molecular asymmetry. In Strasbourg, Pasteur had the immense good fortune to meet and marry the University Rector's daughter Marie Laurent, who was to be his devoted wife, mother and scientific helpmate through the remainder of his life.

In 1854 Pasteur was appointed Dean and professor of chemistry at the Faculty of Sciences in Lille, France. Lille was an industrial town with a number of distilleries and factories. The Minister of Public Instruction was not completely sold on "science for science's sake". He reminded university faculty that (and here I quote the Minister's words) "whilst keeping up with scientific theory, you should, in order to produce useful and far reaching results, appropriate to yourselves the special applications suitable to the real wants of the surrounding country."

Pasteur, in contrast to other faculty, needed no prodding. He enjoyed taking his students on tours of the factories and was quick to advise the managers that he was available to help solve their problems. In the summer of 1856, M. Bigot, father of one of his students in chemistry, called upon Pasteur to help him overcome difficulties he was having manufacturing alcohol by fermentation of beetroot. Often, instead of alcohol, Bigot's fermentations yielded lactic acid.

To better appreciate the discoveries to follow, we should understand what was believed at that time about alcoholic fermentation. Chemistry was emerging as a true science, freed from the pseudoscience of the alchemist. The mysterious chemical processes of living animals were slowly being unraveled in strictly chemical terms. Lavoisier had shown that chemical combustion in living animals was quantitatively identical to that occurring in a furnace. Lavoisier also showed that sugar, the starting product of fermentation, could be broken down to alcohol, CO_2 and H_2O by simply dropping a sugar solution on heated platinum. Woehler startled the scientific world by sythesizing the organic compound urea, showing for the first time that organic compounds, believed up to then as capable of synthesis only by living animals could be made in a test tube. And due, in no small part to Pasteur's work on crystals, internal structure and analysis of complex organic compounds was becoming routine.

In this light, fermentation leading to production of wine, beer and vinegar was believed to be a straightforward chemical breakdown of sugar to the desired molecules. The chemical experts of the day proclaimed that the breakdown of sugar into alcohol during fermentation of sugar to wine and beer was due to the presence of inherent unstabilizing vibrations. One could transfer these unstabilizing vibrations from a vat of finished wine to new grape pressings to start fermentation anew.

Yeast cells were found in the fermenting vats of wine, and were recognized as being live organisms, but they were believed simply to be either a product of fermentation or catalytic agents that provided useful ingredients for fermentation to proceed. Those few biologists who earlier concluded that yeast was the cause of, and not the product of, fermentation were ridiculed by the scientific experts: The deep conviction of the scientific establishment was that chemistry had come too far to allow a vitalistic life force theory to challenge pure chemical explanations of molecular reaction. To attribute such chemical changes to mysterious life forces would represent a major backward step in science!

Unfortunately, the "scientific establishment" was not providing much help to the brewers of wine, beer and vinegar. These manufacturers were plagued by serious economic problems related to their fermentations. Yields of alcohol might suddenly fall off; wine might unexpectedly grow ropy or sour or turn to vinegar; vinegar, when desired, might not be formed and lactic acid might appear in its place; the quality and taste of beer might unexpectedly change making quality control a nightmare! All too often the producers would be forced to throw out the resultant batches, start anew, and sadly have no better luck!

Into M. Bigot's factory, microscope in hand, came Pasteur. He quickly found three clues that allowed him to solve the puzzle of alcoholic fermentation. First, when alcohol was produced normally, the yeast cells were plump and budding. But when lactic acid would form instead of alcohol, small rod like microbes were always mixed with the yeast cells. Second, analysis of the batches of alcohol showed that amyl alcohol and other complex organic compounds were being formed during the fermentation. This could not be explained by the simple catalytic breakdown of sugar shown by Lavoisier. Some additional processes must be involved. Third, and this may have been the critical clue to Pasteur, some of these compounds rotated light, that is they were asymmetric. As we said earlier, Pasteur suspected that only living cells produced asymmetrical compounds. He concluded and was able to prove that living cells, the yeast, were responsible for forming alcohol from sugar, and that contaminating micro-organisms turned the fermentations sour!

Over the next several years Pasteur identified and isolated the specific micro-organisms responsible for normal and abnormal fermentations in production of wine, beer, vinegar. He showed that if he heated wine, beer, milk to moderately high temperatures for a few minutes, he could kill living micro-organism and thereby sterilize (pasteurize), the batches and prevent their degradation. If pure cultures of microbes and yeasts were added to sterile mashes uniform, predictable fermentations would follow.

Spontaneous Generation

In the midst of the great excitement and controversy created by Pasteur's research on fermentation, a debate was ongoing in the scientific world on the theory of "spontaneous generation". The idea that beetles, eels, maggots and now microbes could arise spontaneously from putrefying matter was speculated on from Greek and Roman times. And in the 1860's spontaneous generation was still a subject of debate in the exalted French Academy of Sciences. Against the advice of his colleagues, who saw dabbling in this field as thankless and unrewarding, Pasteur entered the fray. Based on his work on fermentation it seemed obvious to him that the sources of yeasts and other micro-organisms that were found during fermentation and putrefaction entered from the outside, for example, on the dust of the air. Pasteur conducted a series of ingenious experiments that destroyed every argument supporting "spontaneous generation". He showed that the skin of grapes towards the beginning of grape harvest was the source of the yeast. Drawing grape juice from under the skin with sterile needles gave juice that would not ferment. Covering the grape arbors with fine cloth or wrapping the grapes with cotton to keep off contaminating dust, gave grapes that would not produce wine. In order to show that dust of the air was the carrier of contamination, he allowed air collected at different altitudes, from sea level to mountain tops, to enter sterilized vessels containing fermentable solutions. The higher the altitude the less the dust in the air and the fewer flasks showed growth.

The experimental design that clinched the argument was the use of the swan-neck flask. In this experiment, fermentable juice was placed in a flask and after sterilization the neck was heated and drawn out as a thin tube taking a gentle downward then upward arc — resembling the neck of a swan. The end of neck was then sealed. As long as it was sealed, the contents remained unchanged. If the flask was opened by nipping off the end of the neck, air entered but dust was trapped on the wet walls of the neck. Under this condition, the fluid would remain forever sterile, showing that air alone could not trigger growth of micro-organisms. If, however, the flask was tipped to allow the sterile liquid to touch the contaminated walls and this liquid was then returned to the broth, growth of micro-organisms immediately began.

In the words of Pasteur "Never will the doctrine of spontaneous generation recover from the mortal blow of this simple experiment. No, there is now no circumstance known in which it can be affirmed that microscopic beings came into the world without germs, without parents similar to themselves."

Silk Worms

As if Pasteur was not busy enough with his studies on fermentation and spontaneous generation, he was asked by the Department of Agriculture to head a commission to see what could be learned about a devastating disease of silkworms that was destroying the French silk industry. Even though Pasteur knew nothing of silkworms and had no idea that they suffered from disease, his research on silkworms forged another link in his "inevitable" chain of discovery.

Now there were at least two different types of silkworm diseases that Pasteur came to grips with: Pebrine, in which black spots and corpuscles are generally, but not always, present on the worm. In such cases the worms often die within the cocoons. In the second type of disease, flacherie, the worms exhibit no corpuscles or spots but fail to spin cocoons. Pasteur suspected, but was not sure, that pebrine corpuscles were associated with the failure of the worms. Nonetheless, by examining the worms under the microscope he was able to identify those free of pebrine and used only their eggs for breeding. Next he excluded from breeding eggs from worms with flacherie whom he identified by their sluggish behaviour in climbing leaves when about to construct cocoons. He instructed the silkworm farmers on these methods of selection and how to use the microscope to detect sickness in the worms. Soon the silk industry in France, Italy and other European countries returned to health.

Pasteur considered these studies important landmarks in his investigations on infection and infectious disease. As he expanded his research, he found that healthy worms became infected when allowed to nest on leaves used by infected worms. He also noted that the susceptibility of the worms varied widely, some worms dying shortly after infection, some weeks later, some not at all. He determined that temperature, humidity, ventilation, quality of the food, sanitation and adequate separation of the broods of newly hatched worms each played a role in susceptibility to the disease. So here from Pasteur's research we see the emergence of his future concepts of the influence of environment on contagion.

Germ Theory of Disease

The crowning achievements of Pasteur's career were development of the germ theory of disease and the use of vaccines to prevent these diseases. Pasteur's studies on contamination of wine and beer by airborne yeast clearly stimulated certain investigators to recognize that these "diseases" were due to entry of foreign micro-organisms. Lister in England was so impressed by Pasteur's work that he began to systematically sterilize his instruments, bandages and sprayed phenol solutions in his laboratory thus reducing infections following surgery to incredibly low numbers.

By 1875 many physicians recognized that some diseases were accompanied by specific micro-organisms, but the body of medical opinion was unwilling to concede that important diseases—cholera, diphtheria, scarlet fever, childbirth fever, syphilis, smallpox - could ever be caused by these agents. To give you an idea of the magnitude of the problem, according to Pasteur's biographer son-in-law Vallery-Radot between April 1 and May 10, 1856, in the Paris Maternity Hospital there were 64 fatalities due to childbirth fever out of 347 confinements. The hospital was closed and the patients were transferred to a different hospital. Sadly, the contagion followed these women and nearly all of them died!

As Pasteur wandered through hospital wards he became increasingly aware that infection was spread by physicians and hospital attendants from sick to healthy patients. Pasteur impressed on his physician colleagues that avoidance of microbes meant avoidance of infection. In a famous speech before the august Academy of Medicine in Paris he stated, "This water, this sponge, this lint with which you wash or cover a wound, may deposit germs which have the power of multiplying rapidly within the tissue....If I had the honour of being a surgeon....not only would I use none but perfectly clean instruments, but I would clean my hands with the greatest care...I would use only lint, bandages and sponges previously exposed to a temperature of 1300 to 1500 degrees. Slowly, but surely, through the preachings of Pasteur, Lister and other physicians antiseptic medicine and surgery became the rule.

Anthrax

At this time, anthrax, a fatal disease of sheep and cattle, was decimating the sheep industry and the economy of France.

Important strides in identifying the causative agent of anthrax had been made by the time Pasteur entered the arena. The great German physician/scientist Robert Koch, isolated the anthrax bacillus, previously identified by the French physician Davain, from infected spleens and showed that under resting conditions the bacillus formed long-lived spores.

Definitive proof was still lacking that the cultured bacillus, itself, and not something carried along in Koch's culture medium was responsible to giving injected animals anthrax. Pasteur provided this proof. As described by Dubos, Pasteur placed one drop of blood from a sheep dying of anthrax into 50 ml of sterile culture, grew up the bacterium, and then repeated this process 100 times. This represented a huge dilution of the original culture so that not a single molecule of the original culture remained in the final culture. Yet, the last culture was as active as the first in producing anthrax. As only the bacillus, itself, by growing up each time in the new culture, could escape dilution, it proved beyond all doubt that the anthrax bacillus and nothing else could be responsible for the disease. Thus was the germ theory of disease firmly established!

But how did the disease spread? Why was one field deadly to sheep, another harmless? Here Pasteur's studies on silkworm contagion provided the clue. During one of Pasteur's excursions to a field where sheep were grazing he noted that the ground in one part of the field was differently coloured than the rest. There it was that the farmer had buried some sheep dead of anthrax. The colour of the soil was due to earth worm casts. He realized that earth worms were feeding on the carcasses of the buried sheep and bringing the anthrax spores to the surface where other sheep could graze on the contaminated soil. Although containment of the animals on uncontaminated fields would help control the spread of anthrax, more was needed.

Interestingly, Pasteur's studies on chicken cholera going on at this time provided the breakthrough that led to development of specific vaccines to fight disease. Cholera was a serious problem for farmers. Chicken cholera would spread through a barnyard rapidly and wipe out the entire flock in as little as 3 days. Spread could be by contaminated food or animal excrements. Pasteur had identified the cholera bacillus and was growing it in pure culture. When injected, chicken invariably died in 48 hours.

Then luck intervened. During the heat of the summer, Pasteur returned to Paris leaving the cholera cultures used for infection stored on the shelves of the Arbois laboratory. Upon return, Pasteur's collaborators were disappointed to find that these stored cultures no longer killed injected chickens, nor even made them sick. The group set to work to make new cultures of the bacillus and tested these batches on new birds and those healthy previously treated birds. The results were astonishing: The previously injected birds were unaffected by the bacillus, while the new birds all died. When Pasteur saw these results he immediately realized that in a sense he was repeating the studies of Jenner 80 years earlier who had conferred on humans immunity to smallpox by vaccinating individuals with a mild form of cowpox. Pasteur then reproducibly manufactured attenuated cultures of chicken cholera vaccines and could routinely prevent cholera in the vaccinated chickens.

If attenuated cholera bacillus could render chickens resistant to the disease, would not an attenuated anthrax bacillus render sheep immune to anthrax? By various techniques involving oxidation and aging, anthrax vaccines indeed prevented anthrax in laboratory trials. Pasteur's reports on preventing sheep anthrax were so exciting to some and unbelievable to many, that he was challenged by the well-known veterinarian Rossignol to conduct a carefully controlled public test of his anthrax vaccine. This was to take place at Pouilly le Fort, a farm in the town of Melun south of Paris. Twenty-five sheep were to be controls, the other twenty-five were to be vaccinated by Pasteur and then all animals would receive a lethal dose of anthrax. All of the control sheep must die and the vaccinated sheep must live. When Pasteur's colleagues learned that he had agreed to the test they were concerned. The challenge was severe and there was no room for error. The vaccines were still in the developmental stage. "What succeeded with 14 sheep in our laboratory will succeed with 50 at Melun", said Pasteur.

The publicity was intense. A reporter from the London Times sent back daily dispatches. Newspapers in France followed the events with daily bulletins. There were crowds of onlookers, farmers, engineers, veterinarians, physicians, scientists and a carnival atmosphere. Would Pasteur's claims of vaccination hold up? Even Pasteur was privately concerned that he had acted impetuously in accepting the challenge. Happily, the trial was a complete success — indeed, a triumph! Two days after final inoculation (May 5, 1882), every one of 25 control sheep was dead and every one of the 25 vaccinated sheep was alive and healthy. The fame of Pasteur and these experiments spread throughout France, Europe and beyond. It was, says Duclaux, "the anthrax vaccine that spread through the public mind faith in the science of microbes". Within 10 years a total of 3.5 million sheep and a half million cattle had been vaccinated with a mortality of less than 1%. The immediate savings to the French economy were enormous, at least 7 million francs, estimated to be enough to cover the reparations that France was required to pay to Prussia for the loss of the Franco-Prussian War in 1880.

Supported by the successes with anthrax and fowl cholera diseases, Pasteur identified and isolated over the next 2-3 years the microbes for many other diseases including swine erysipelas, childbirth fever and pneumonia.

Rabies

The final and certainly most famous success of Pasteur's research was the development of a vaccine against rabies or hydrophobia as it is also known. The disease has always had a hold on the public imagination and has been looked upon with horror. It evokes visions of "raging victims, bound and howling, or asphyxiated between two mattresses" (Duclaux). The treatments applied to victims were as horrible as the supposed symptoms: this included cauterizing the bite wounds with a red-hot poker. Actually very few persons die in any year from being bitten by a rabid dog or wolf. The symptoms of the disease are variable: onset may take weeks to months to develop if they develop at all. Nonetheless, Pasteur and his colleague Roux realized that conquest of rabies would be recognized as a great achievement to the world of science and to the public at large.

Pasteur and Roux initially attempted to transfer infection by injecting healthy dogs with saliva from rabid animals. The results were variable and unpredictable. Later, recognizing that the active agent was in the spinal cord and brain, and because they were unable to detect a specific rabic micro-organism, Pasteur and Roux applied extracts of rabid spinal cord directly to the brain of dogs. With this technique they could reproducibly produce rabies in the test animals in a few days.

The goal was next to develop a vaccine that would provide protection to the subject before the rabic agent moved from the bite site to the spinal cord to the brain. This was achieved by injecting into test animals suspensions of spinal cord of rabid rabbits that were attenuated in strength by air drying over a 12-day period in the now-famous Roux Bottle. A strip of spinal cord was suspended from a hanger in the center of the bottle containing a hole at the top of the bottle and one on the lower side. Air entered from the bottom opening, passed over a drying agent and exited from the top. The longer the cord was dried, the less potent was the tissue in producing rabies.

The treatment plan used to develop immunity to rabies was to inject under skin of a dog the least potent preparation of minced spinal cord, followed every day for the next 12 days with a stronger and stronger extract. At the end of this time, the animal was completely resistant to bites of rabid dogs and failed to develop rabies if the most potent extracts were applied directly to the brain.

Following confirmation of his reports in 1885 that he had made dogs refractory to rabies by vaccination, Pasteur received wide acclaim and much favourable publicity. But why not use the vaccine on humans? Frankly, Pasteur was terribly afraid of things going wrong and he was particularly uneasy about being unable to isolate the rabic substance. And so he continued to insist that many years of additional research was necessary before the treatment could be tried on humans.

But the press of events made him act sooner. On July 6, 1886, 9 year old Joseph Meister and his mother appeared at Pasteur's laboratory. Two days earlier the young boy had been bitten repeatedly by a rabid dog. He was so badly mauled that he could hardly walk. His mother appealed to Pasteur to treat her son. At the time Pasteur had treated about 40 dogs, most of whom resisted rabies. Could he risk treating this youth who faced certain death? Pasteur, after consultation with physician colleagues, and much trepidation treated the youth. Despite Pasteur's fears, Meister made a perfect recovery and remained in fine health for the remainder of his life.

A few months later a second victim turned up. He was a young shepherd also bitten by a mad dog. Following reports of his successful treatments, the wild acclaim for Pasteur knew no bounds! Victims of dog and wolf bites from France, Russia, the United States poured into his laboratory for treatment. The newspapers and public followed these treatments and cures with intense interest. Pasteur became a hero and a legend. The Pasteur Institute funded by public and governmental subscriptions was built in Paris initially to treat victims of rabies who were coming to Pasteur's laboratory in increasing numbers. Later, Pasteur Institutes were built, including 3 in the United States, to deal with human rabies and other diseases.

Rabies was the last major research of the master scientist. His health was failing and a paralysis of his left side from a serious stroke he suffered in his 46th year made his working in the laboratory increasingly difficult. Pasteur died in 1895 after suffering additional strokes. He was buried, a national hero, by the French Government. His funeral was attended by thousands of people. His remains, initially interred in the Cathedral of Notre Dame, was transferred to a permanent crypt in the Pasteur Institute, Paris.

In a tragic footnote to history, Joseph Meister, the first person publicly to receive the rabies vaccine, returned to the Pasteur Institute as an employee where he served for many years as Gatekeeper. In 1940, 45 years after his treatment for rabies that made medical history, he was ordered by the German occupiers of Paris to open Pasteur's crypt. Rather than comply, Joseph Meister committed suicide!

ALFRED NOBEL

Alfred Nobel was born in Stockholm on October 21, 1833. His father Immanuel Nobel was an engineer and inventor who built bridges and buildings in Stockholm. In connection with his construction work Immanuel Nobel also experimented with different techniques for blasting rocks.

Alfred's mother, Andriette Ahlsell, came from a wealthy family. Due to misfortunes in his construction work caused by the loss of some barges of building material, Immanuel Nobel was forced into bankruptcy the same year Alfred Nobel was born. In 1837 Immanuel Nobel left Stockholm and his family to start a new career in Finland and in Russia. To support the family, Andriette Nobel started a grocery store which provided a modest income. Meanwhile Immanuel Nobel was successful in his new enterprise in St. Petersburg, Russia. He started a mechanical workshop which provided equipment for the Russian army and he also convinced the Tsar and his generals that naval mines could be used to block enemy naval ships from threatening the city.

The naval mines designed by Immanuel Nobel were simple devices consisting of submerged wooden casks filled with gunpowder. Anchored below the surface of the Gulf of Finland, they effectively deterred the British Royal Navy from moving into firing range of St. Petersburg during the Crimean war (1853-1856). Immanuel Nobel was also a pioneer in arms manufacture and in designing steam engines.

Successful in his industrial and business ventures, Immanuel Nobel was able, in 1842, to bring his family to St. Petersburg. There, his sons were given a first class education by private teachers. The training included natural sciences, languages and literature. By the age of 17 Alfred Nobel was fluent in Swedish, Russian, French, English and German. His primary interests were in English literature and poetry as well as in chemistry and physics. Alfred's father, who wanted his sons to join his enterprise as engineers, disliked Alfred's interest in poetry and found his son rather introverted. In order to widen Alfred's horizons his father sent him abroad for further training in chemical engineering. During a two year period Alfred Nobel visited Sweden, Germany, France and the United States. In Paris, the city he came to like best, he worked in the private laboratory of Professor T. J. Pelouze, a famous chemist. There he met the young Italian chemist Ascanio Sobrero who, three years earlier, had invented nitroglycerine, a highly explosive liquid. Nitroglycerine was produced by mixing glycerine with sulphuric and nitric acid. It was considered too dangerous to be of any practical use. Although its explosive power greatly exceeded that of gunpowder, the liquid would explode in a very unpredictable manner if subjected to heat and pressure. Alfred Nobel became very interested in nitroglycerine and how it could be put to practical use in construction work. He also realized that the safety problems had to be solved and a method had to be developed for the controlled detonation of nitroglycerine. In the United States he visited John Ericsson, the Swedish-American engineer who had developed the screw propeller for ships. In 1852 Alfred Nobel was asked to come back and work in the family enterprise which was booming because of its deliveries to the Russian army. Together with his father he performed experiments to develop nitroglycerine as a commercially and technically useful explosive. As the war ended and conditions changed, Immanuel Nobel was again forced into bankruptcy. Immanuel and two of his sons, Alfred and Emil, left St. Petersburg together and returned to Stockholm. His other two sons, Robert and Ludvig, remained in St. Petersburg. With some difficulties they managed to salvage the family enterprise and then went on to develop the oil industry in the southern part of the Russian empire. They were very successful and became some of the wealthiest persons of their time.

After his return to Sweden in 1863, Alfred Nobel concentrated on developing nitroglycerine as an explosive. Several explosions, including one (1864) in which his brother Emil and several other persons were killed, convinced the authorities that nitroglycerine production was exceedingly dangerous. They forbade further experimentation with nitroglycerine within the Stockholm city limits and Alfred Nobel had to move his experimentation to a barge anchored on Lake Mälaren. Alfred was not discouraged and in 1864 he was able to start mass production of nitroglycerine. To make the handling of nitroglycerine safer Alfred Nobel experimented with different additives. He soon found that mixing nitroglycerine with silica would turn the liquid into a paste which could be shaped into rods of a size and form suitable for insertion into drilling holes. In 1867 he patented this material under the name of dynamite. To be able to detonate the dynamite rods he also invented a detonator (blasting cap) which could be ignited by lighting a fuse. These inventions were made at the same time as the diamond drilling crown and the pneumatic drill came into general use. Together these inventions drastically reduced the cost of blasting rock, drilling tunnels, building canals and many other forms of construction work.

The market for dynamite and detonating caps grew very rapidly and Alfred Nobel also proved himself to be a very skillful entrepreneur and businessman. By 1865 his factory in Krümmel near Hamburg, Germany, was exporting nitroglycerine explosives to other countries in Europe, America and Australia. Over the years he founded factories and laboratories in some 90 different places in more than 20 countries. Although he lived in Paris much of his life, he was constantly travelling. Victor Hugo at one time described him as "Europe's richest vagabond". When he was not travelling or engaging in business activities Nobel himself worked intensively in his various laboratories, first in Stockholm and later in Hamburg (Germany), Ardeer (Scotland), Paris and Sevran (France), Karlskoga (Sweden) and San Remo (Italy). He focused on the development of explosives technology as well as other chemical inventions, including such materials as synthetic rubber and leather, artificial silk, etc. By the time of his death in 1896 he had 355 patents.

Intensive work and travel did not leave much time for a private life. At the age of 43 he was feeling like an old man. At this time he advertised in a newspaper "Wealthy, highly-educated elderly gentleman seeks lady of mature age, versed in languages, as secretary and supervisor of household." The most qualified applicant turned out to be an Austrian woman, Countess Bertha Kinsky. After working a very short time for Nobel she decided to return to Austria to marry Count Arthur von Suttner. In spite of this Alfred Nobel and Bertha von Suttner remained friends and kept writing letters to each other for decades. Over the years Bertha von Suttner became increasingly critical of the arms race. She wrote a famous book, *Lay Down Your Arms* and became a prominent figure in the peace movement. No doubt this influenced Alfred Nobel when he wrote his final will which was to include a Prize for persons or organizations who promoted peace. Several years after the death of Alfred Nobel, the Norwegian Storting (Parliament) decided to award the 1905 Nobel Peace Prize to Bertha von Suttner.

Alfred Nobel's greatness lay in his ability to combine the penetrating mind of the scientist and inventor with the forward-looking dynamism of the industrialist. Nobel was very interested in social and peace-related issues and held what were considered radical views in his era. He had a great interest in literature and wrote his own poetry and dramatic works. The Nobel Prizes became an extension and a fulfillment of his lifetime interests.

Many of the companies founded by Nobel have developed into industrial enterprises that still play a prominent role in the world economy, for example Imperial Chemical Industries (ICI), Great Britain; Société Centrale de Dynamite, France; and Dyno Industries in Norway. Toward the end of his life, he acquired the company AB Bofors in Karlskoga, where Björkborn Manor became his Swedish home. Alfred Nobel died in San Remo, Italy, on December 10, 1896. When his will was opened it came as a surprise that his fortune was to be used for Prizes in Physics, Chemistry, Physiology or Medicine, Literature and Peace. The executors of his will were two young engineers, Ragnar Sohlman and Rudolf Lilljequist. They set about forming the Nobel Foundation as an organization to take care of the financial assets left by Nobel for this purpose and to coordinate the work of the Prize-Awarding Institutions. This was not without its difficulties since the will was contested by relatives and questioned by authorities in various countries.

ELISHA GRAVES OTIS

Elisha Graves Otis was born on August 3, 1811. Though widely believed to be inventor of the elevator, he did in fact invent something perhaps more important-the elevator brake-which made skyscrapers a practical reality.

Invention Impact

Otis had no way of knowing that this simple safety device was to alter the face of the globe, that because of it vast cities would spring up toward the sky instead of spreading toward the horizon as in the past.

Inventor Bio

Born on a farm near Halifax, Vermont, the youngest of six children, Otis made several attempts at establishing businesses in his early years. However, chronically poor health led to continual financial woes.

Finally, in 1845, he tried to change his luck with a move to Albany, New York. There he worked as a master mechanic in the bedstead factory of O. Tingley & Company. He remained about three years and during that time invented and put into use a railway safety brake, which could be controlled by the engineer, and ingenious devices to run rails for four-poster beds and to improve the operation of turbine wheels.

By 1852 he had moved to Yonkers, New York, to organize and install machinery for the bedstead firm of Maize & Burns, which was expanding. Josiah Maize needed a hoist to lift heavy equipment to the upper floor. Although hoists were not new, Otis' inventive nature had been piqued because of the equipment's safety problem.

If one could just devise a machine that wouldn't fall.... He hit upon the answer, a tough, steel wagon spring meshing with a ratchet. If the rope gave way, the spring would catch and hold.

In 1854 Otis dramatized his safety device on the floor of the Crystal Palace Exposition in New York. With a large audience on hand, the inventor ascended in an elevator cradled in an open-sided shaft. Halfway up, he had the hoisting cable cut with an axe. The platform held fast and the elevator industry was on its way.

ALEXANDER GRAHAM BELL

Alexander Graham Bell (Mar. 3, 1847-Aug. 2, 1922), inventor of the telephone and the outstanding figure of his generation in the education of the deaf, was born in Edinburgh, the second of the three sons of Alexander Melville Bell, a scientist and author in the field of vocal physiology and elocution, and of Eliza Grace Symonds, the daughter of a surgeon in the Royal Navy. His grandfather, Alexander Bell (1790-1865), was a professor of elocution in London. From both sides of the family he inherited a scientific tendency and an instinct for applying knowledge to practise. During his earlier years he was taught at home, by his mother, a woman of admirable character who was also unusually gifted, a musician and painter of ability. From her Bell obtained his accuracy and sensitiveness of hearing and his love for music. When he was ten years old he went to McLauren's Academy in Edinburgh, and then to the Royal High School, from which he graduated at the age of thirteen. A visit to his grandfather in London extended to more than a year, and brought him many educational and cultural advantages. At the age of sixteen he secured a position as a pupil-teacher of elocution and music in Weston House Academy, at Elgin in Morayshire. The next year he spent at the University of Edinburgh, and then returned to Elgin for two years, thereafter spending a year (1866-67) as an instructor in Somersetshire College at Bath in England.

The distinctive achievement of his father, Alexander Melville Bell, was the invention of Visible Speech, a system of symbols by which the position of the vocal organs in speech was indicated. When the grandfather died in 1865, Alexander Melville Bell moved to London to take up his father's work there, and about two years later Alexander Graham Bell became his father's professional assistant in London. At this time he matriculated at University College, London, taking courses in anatomy and physiology for three years, 1868-70. While his father was absent in America on a lecture tour in 1868, Alexander Graham Bell took entire charge of his father's professional duties. The next year Alexander Melville Bell took his son into full partnership with him, an arrangement which continued until the family sailed for America, arriving in Quebec on Aug. 1, 1870.

Alexander Graham Bell early showed an aptness for systematic investigation and promise of inventive ability, which his father encouraged. The former's first piece of original scientific work was set forth in a letter to his father, from Elgin, Nov. 24, 1865, on the resonance pitches of the mouth cavities during the utterance of vowel sounds. His father suggested to him that he send this study to Alexander J. Ellis, an eminent phonetician and friend in London. Ellis called his attention to the work of Helmholtz on vowel sounds. This resulted in Bell's beginning the study of electricity as applied in telegraphy. At this time he conceived his first idea for the electrical transmission of speech. His work for the deaf, also, grew out of his position as his father's assistant. One of his father's pupils, Miss Susanna E. Hull, had a school for deaf children in Kensington. At her request in May 1868, Alexander Melville Bell sent his son, then twenty-one years old, to her school to adapt Visible Speech and its teaching to the work for the deaf, and during the same year in an influential lecture at the Lowell Institute, Boston, he told of the use of Visible Speech in the London School. As a result, in the following year through the efforts of the Rev. Dexter King a special day school for the deaf, the first of its kind anywhere, was started by the Boston School Board with Miss Sarah Fuller as principal. She had heard Alexander Melville Bell's lecture and persuaded the School Board to engage his son to come and train her teachers in the use of Visible Speech. Bell began his work at Miss Fuller's school on Apr. 10, 1871. He added to Visible Speech a system of notation that improved it for use with the deaf. This is still the basis of the method used in teaching the deaf to talk.

After three months of Miss Fuller's school, Bell visited the Clarke Institution for the Deaf at Northampton, Mass and the American Asylum for the Deaf at Hartford, Conn. As the demand for his services increased, in October 1872 he opened a private normal class in Boston to which institutions could send teachers for training in the use of Visible Speech. Early in 1873 he was appointed professor of vocal physiology and the mechanics of speech in the School of Oratory of Boston University. There also he started a normal class for the training of teachers of the deaf. Bell's lectures at Boston University attracted wide attention and the University of Oxford invited him to deliver a course there. This he did with notable success in 1878. His normal work became still more extensive through a series of conventions of teachers of speech to the deaf, which he started and led. His work also became more intensive, as he took a number of private pupils. One of these was the five-year-old son of Thomas Sanders of Haverhill, Mass who was born deaf. Bell had charge of the child's entire education for more than three years, 1873-76, living with him at the home of the grandmother in Salem. Believing that the principle of Froebel's kindergarten method would be useful in the teaching of the deaf, he used both playing and working as direct means of instruction.

Bell's inventive activities were not suspended by these interests, though invention was always subordinate in his own mind to his main purpose to devote his life to the welfare of the deaf. His approach to his inventive work through the science of acoustics and his work for the deaf was particularly favourable to the success later of his telephonic experiments. Further, so remarkable was his teaching of the little Sanders boy that Thomas Sanders in generous gratitude offered to meet all the expenses of his experimenting and of securing patents for his inventions. A little later Gardiner G. Hubbard, a man of notable public spirit and an active friend of the education of the deaf, also became interested in Bell and gave most important and effective assistance in the commercial management of his inventions. During the years 1873-76, Bell was experimenting along three related lines—to invent a phonautograph; a multiple telegraph;

and an electric speaking telegraph or telephone. He hoped in his phonautograph to invent an instrument with which he could explain to his deaf pupils how to make their tone-vibrations correctly by comparing visual records of the sounds they made with standard records. For this Dr. Clarence J. Blake, an aurist, suggested that he use an actual human ear in studying the question, and prepared one for him. Bell never brought his phonautograph to such a stage as to be of practical use for his purpose, but it was of value to him in his telephone experiments; from it he took the conception of the membrane element in the telephone.

His experiments in telegraphy Bell had begun at Elgin when only eighteen. At that time telegraphy constituted the whole art of electrical communication. Before leaving London for America in 1870, Bell had secured a copy in French of Helmholtz's book, *The Sensations of Tone.* His study of this book was an important help in gaining a correct knowledge of the physical principles underlying the theory of sound. Thereafter he advanced with a firmer step in all his researches. After coming to America, what time he could spare from his work for the deaf he gave to telegraphy, hoping to invent a device that would send two or more telegraph messages over a wire at the same time. In his multiple harmonic telegraph he utilized the fact that a tuned reed will vibrate when its own note is sounded near it. A number of reed transmitters were attached by leads to the one wire; at the other end of the wire a similar number of leads ran to receivers in which the reeds were attuned accurately to their corresponding transmitters. Each receiver would in operation vibrate in response only to its own transmitter. These experiments resulted in two patents. The first was No. 161,739, Apr. 6, 1875, for an Improvement in Transmitters and Receivers for Electrical Telegraphs; the other was No. 178,399, June 6, 1876, for Telephonic Telegraph Receivers. From his arduous study and experimentation in telegraphy Bell gained invaluable advantage in the mastery of the principles of electrical wave transmission.

But the speaking telephone was always in Bell's mind the most important invention. In the summer of 1874, following his usual custom, he had gone to his father's home at Brantford, Ontario, to spend his vacation. While there, on July 26, 1874, Bell tells us, he formed clearly in his mind the theory of the telephone that ultimately proved to be correct. How to realize this theory in practise was a problem which presented many difficulties. Bell had learned at the Massachusetts Institute of Technology that some of the points he had discovered concerning the application of acoustics to telegraphy had already been discovered by Joseph Henry [*q. v.*]. In order to learn which of his own discoveries were new and which were old he called upon Prof. Henry at the Smithsonian Institution. The aged scientist treated him with genuine interest and gracious respect, even asking permission to verify Bell's experiments and publish them through the Smithsonian Institution, giving Bell full credit for them. Much encouraged, Bell told Joseph Henry about his telephone idea and asked Henry whether he ought to publish it at its present stage and allow others to work it out or try to finish it himself. Henry declared it to be the "germ of a great invention" and told him to go ahead himself. Bell then said that he lacked the electrical knowledge that was necessary. Henry emphatically replied, "Get it!" This sympathetic and unhesitating confidence in him by Joseph Henry led Bell to persevere. Whatever observations he made, whatever experiments he was conducting, his mind henceforth was always alert to note anything which might throw light on his telephone problem. At length on June 2, 1875, in the shop at 109 Court St. Boston, while preparing the apparatus used in his harmonic telegraph for an experiment, Bell observed an effect, the significance of which he alone was competent to appreciate. He was tuning the receiver reeds and his assistant, Thomas A. Watson, was plucking the transmitting reeds to give him the pitch over the wire. In one transmitter the contact point was screwed down too far and Bell's sensitively accurate ear heard not only the pitch or tone of the reed but also the overtones. He knew at once that he had found what he had been seeking, that any condition of apparatus that would reproduce the tone and overtones of a steel spring could be made to reproduce the tone and overtones of the human voice. After repeatedly verifying the fact he had observed, Bell gave Watson directions for making the first telephone. When tested the next morning Watson could recognize Bell's voice, and could almost understand some of the words. Experimenting to improve the quality of the transmission followed. On March 10, 1876, the telephone transmitted its first complete intelligible sentence, "Mr. Watson, come here; I want you." Bell's subsequent development of the telephone in quality and distance of transmission was rapid, until in about a year, April 3, 1877, he was able to conduct a telephone conversation between Boston and New York with some degree of success. In September and October 1875, Bell had written the specifications for his application for a patent.

The historic first Telephone Patent, No. 174,465, had been allowed on Bell's twenty-ninth birthday, March 3, and had been issued to him on March 7, 1876. On January 30, 1877, another patent was issued to him, No. 186,787, for the substitution of an iron or steel diaphragm for the membrane and armature in the telephone of the first patent. Many claimants now came forward to contest Bell's rights, but after the most prolonged and important litigation in the history of American patent law, including about 600 cases, the United States Supreme Court upheld all of Bell's claims, declaring that he was the discoverer of the only way that speech could be transmitted electrically (*126 United States Reports*). His first announcement and demonstration of the telephone to the scientific world was in an address before the American Academy of Arts and Sciences in Boston on May 10, 1876. On June 25, 1876, at the International Centennial Exhibition at Philadelphia, through the interest of Dom Pedro II, the Emperor of Brazil, who was present as a guest of honour, and to whom a few months before Bell had shown the school for the deaf in Boston, Bell had an opportunity to explain his telephone to the judges, among whom were Sir William Thomson, Joseph Henry, and other prominent scientists. This demonstration was in itself an important introduction of the telephone to the scientific world. Lectures before scientific societies led to a demand for lectures for the general public.

At Salem on Feb. 12, 1877, the telephone was used for the first time to send a report to a newspaper. With the spring of 1877 the commercial development of the invention came to the front. The first telephone organization was effected on July 9, 1877, in the form of a trusteeship called the Bell Telephone Company with Gardiner G. Hubbard as trustee and Thomas Sanders as treasurer.

On July 11, 1877, Bell married Mabel G. Hubbard, the daughter of Gardiner G. Hubbard. This girl of eighteen, who was to be an exceptional helpmate for the rest of his life, was entirely deaf from early childhood. There was, therefore, a keen personal element in his sympathy for the deaf and in his work for them. Early in August he sailed with his bride for Europe to introduce the telephone into England and France. A prospectus which he wrote for a group of capitalists in London contains a remarkable prediction of the development of telephone service, describing in detail its uses fifty years later. While he was in England he also gave considerable attention to the education of the deaf, and took part in the inception of a day school for the deaf at Greenock in Scotland.

He returned to America in the fall of 1878, and that winter moved to Washington, D.C. His inventive work now followed related lines rather than the development of the telephone itself. In 1880 the French Government awarded him the Volta Prize of 50,000 francs, for the invention of the electric speaking telephone. He devoted this money to the promotion of research and invention and to the work for the deaf by financing the Volta Laboratory, in which he associated with him his cousin, Chichester A. Bell, and Sumner Tainter. Each had his own special work, and all three cooperated on some undertaking of common interest and probable profit whereby the laboratory might be maintained. Bell continued in the Volta Laboratory an investigation, which he had started in 1878, into the causes and conditions of congenital deafness in New England. He prepared a memoir on *The Formation of a Deaf Variety of the Human*

Race (1884) and a study of the results of the marriage of the deaf.

He also initiated a movement for certain fundamental improvements in the taking of the Census of 1890 with respect to the deaf. During these years of the Volta Laboratory he invented the photophone, an apparatus for transmitting speech over a ray of light by means of the variable electric resistance of selenium to light and shade. He also invented the induction balance for locating metallic objects in the human body, first used on President Garfield, and the telephone probe which he developed from the former invention. These he did not patent but gave to the world. He further invented an audiometer, thus bringing his inventive work a step nearer to his service to the deaf.

Bell was interested in Edison's invention for recording sound, the phonograph. But Edison's tinfoil records left much to be desired. Accordingly the members of the Volta Laboratory Association invented an improved recorder, a flat wax record, a wax cylinder record, and an improved reproducer, and jointly received patents for these improvements on May 4, 1886. When these patents were sold to the American Gramophone Company, the Laboratory was converted into a Volta Bureau for the Increase and Diffusion of Knowledge Relating to the Deaf. This Bureau worked in close cooperation with the American Association for the Promotion of the Teaching of Speech to the Deaf, organized in 1890, of which Bell was elected president and to which he gave all together more than $300,000.

Through his study of marriage among the deaf, Bell was led to give attention to the whole field of longevity and eugenics. An important contribution to this subject was his *Duration of Life and Condition Associated with Longevity* (1918). In this connection he made an excursion into the breeding of multi-nippled and twin-bearing sheep at his summer home on Cape Breton Island. This broad interest in eugenics and his optimistic attitude toward its problems was recognized by his election as honourary president of the Second International Congress of Eugenics.

During the last twenty-five years of Bell's life aviation was his predominant interest. He was one of the first to consider aerial locomotion practicable, and in 1891 encouraged and financially cooperated with Samuel P. Langley, the secretary of the Smithsonian Institution, in the study of aviation. His own experiments followed the line of kite development and resulted in his invention of the tetrahedral kite and the application of that principle of construction to various uses. In 1907 Bell founded the Aerial Experiment Association, of which he was president and to which he gave $50,000. The first public flight of a heavier-than-air machine was made under the auspices of this organization in 1908. He and his associates also solved the problem of the stability of balance in a flying machine, using rigid instead of flexible supporting surfaces and providing ailerons for the wings and the rudders. During the war Bell invented a motorboat which attained a speed of seventy-one miles an hour. In 1883 with the cooperation of Gardiner G. Hubbard, he established *Science,* now the organ of the American Association for the Advancement of Science, and maintained it for several years. From 1896 to 1904 he was president of the National Geographic Society and did much to forward the development of the society and its magazine.

In 1898 he was appointed a regent of the Smithsonian Institution, serving continuously until his death. In 1891 he had started by a generous gift the Astrophysical Observatory of the Smithsonian, and in 1904 he brought the body of James Smithson from Genoa to Washington. The honours, medals, and degrees that were received by him were numerous. In these marks of recognition from Harvard, Oxford, Heidelberg, Edinburgh, and other institutions, Bell's work for the deaf, his inventions for the hearing, and his other services and attainments were alike honoured. In 1915 he opened the first transcontinental telephone line from New York to San Francisco. In 1917, at Brantford, Ontario, the Duke of Devonshire, as governor general of Canada, unveiled the Bell Telephone Memorial in honour of the inventor of the electric speaking telephone and dedicated the old home at Tutelo Heights as a public park.

In 1920, the home of his childhood, Edinburgh, Scotland, conferred upon him the freedom of the city and elected him a burgess and guild brother of the city. He had taken out his first papers for citizenship in the United States at Lawrence, as early as Oct. 27, 1874, and he received his final papers from the supreme court of the District of Columbia at Washington, Nov. 10, 1882, but he never lost his love for his native Scotland and had a large estate in the new Scotland (Nova Scotia), on the Bras d'Or Lakes of Cape Breton Island. There he spent his summers, and there he died. He was buried on top of a mountain in a tomb cut in the rock, while every telephone on the continent of North America remained silent.

In appearance, during his earlier years, he was a tall, spare young man, with pale complexion, piercing black eyes and bushy, jet black hair and side whiskers, quick of motion, serious and grave. In his later years he was conspicuous for his majestic presence and his radiant manner. His hair and beard had become pure white but his keen black eyes still dominated the situation. The gravity of his youth had given way to a sympathy and joviality which won both old and young.

THOMAS EDISON

Contrary to popular belief, Thomas Edison was not born into poverty in a backwater mid-western town. Actually, he was born (on Feb. 11, 1847) to middle-class parents in the bustling port of Milan, Ohio, a community that - next to Odessa, Russia - was the largest wheat shipping center in the world.

In 1854, his family moved to Port Huron, Michigan, which ultimately surpassed the commercial pre-eminence of both Milan and Odessa. At age seven - after spending 12 weeks in a noisy one-room schoolhouse with 38 other students of all ages - Tom's overworked and short-tempered teacher finally lost his patience with the child's relatively self centered behaviour and persistent questioning. Noting that Tom's forehead was exceptionally broad and his head was larger than average, he made no secret of his belief that the hyperactive youngster's brains were "addled" or scrambled. If modern psychology had existed then, Tom would have probably been deemed a victim of A.D.S. (attention deficit syndrome) and proscribed a hefty dose of the "miracle drug" Ritalin. Instead, when his beloved mother - whom he once said *"... was the making of me and was always so true and so sure of me, I felt I had someone to live for, someone I must not disappoint."* - Convinced her son's slightly unusual physical appearance was a sign of his intelligence. When she became aware of the situation, she promptly withdrew him from school and began to "home-teach" him.

A descendant of the prominent Elliot family of Massachusetts, the devout daughter of a highly respected Presbyterian minister, and an educator in her own right, Nancy Edison (above) now commenced teaching her favourite son the "Three Rs" and the Bible. Meanwhile, his roguish and "worldly" father, Samuel, encouraged him to read the great classics, giving him a ten cents reward for each one he completed. It wasn't long before the serious minded youngster developed a deep interest in world history and English literature. Interestingly, many years later, Tom's abiding fondness for Shakespeare's plays led him to briefly consider becoming an actor. However, because of his high-pitched voice and extreme shyness before every audience - except those he was trying to influence into helping him finance an invention - he soon gave up the idea. Tom especially enjoyed reading and reciting poetry. His life-long favourite was *Gray's Elegy In A Country Churchyard.* Indeed, his favourite lines - which he endlessly chanted to friends, employees, and himself - came from its 9th stanza: *"The boast of heraldry of pomp and power, All that beauty all that wealth ere gave, Alike await the inevitable hour. The path to glory leads but to the grave."* At the age of 11, Tom's parents tried to appease his ever more voracious appetite for knowledge by teaching him how to use the resources of the local library. This was the earliest of many factors that gradually led him to prefer learning through independent self instruction. Starting with the last book on the bottom shelf, Tom began to read what he planned would be every book in the stacks. However, his parents wisely directed him towards being more selective.

By age 12, Tom had not only completed Gibbon's Rise And Fall of the Roman Empire, Sears' History of the World, and Burton's Anatomy of Melancholy, he had also devoured The World Dictionary of Science and a number of works on Practical Chemistry. Unfortunately, in spite of their noble efforts, Tom's dedicated parents found themselves incapable of addressing his ever increasing interest in the Science. For example, when he began to question them about concepts dealing with physics - such as those contained in Isaac Newton's "Principia" - they were utterly stymied. Accordingly, they scraped enough money together to hire a clever tutor to help their precocious son understand Newton's mathematical principles and unique style. Unfortunately, the experience had some negative affects on the highly impressionable boy. Essentially, he was so disillusioned by how Newton's sensational theories were written in classical aristocratic terms - which he felt were unnecessarily confusing to the average person - he overreacted and developed a hearty dislike for all such "high-tone" language and mathematics.

On the other hand, the simple beauty of Newton's *physical* laws did not escape him. They helped him sharpen his *own* free wheeling style of clear and solid thinking, proving *all* things to himself through his own method of objective examination and experimentation." Tom's response to the Principia *also* enhanced his propensity towards gleaning insights from the writings and activities of great men of wisdom, always keeping in mind that even they might be entrenched in preconceived dogma and mired down in associated error.... Meanwhile, Tom cultivated a strong sense of perseverance, readily expending whatever amount of perspiration was needed to meet and overcome all challenges - which was a characteristic he would later note was contrary to how most people respond to challenges. Certainly, his extraordinary mental and physical stamina stood him in good stead when he took on the incredible rigours of a being a successful inventor in the late 19th Century.

Another factor that very much shaped Tom's unique personality was his loss of hearing. Even though this condition - and the fact that he had only three months of formal schooling prevented him from taking advantage of the benefits of a secondary education in contemporary mathematics, physics, and engineering - he never let it interfere with finding ways of compensating. In sum, Tom's "free wheeling" style of acquiring knowledge eventually led him to specifically question many of the prevailing theories on the workings of electricity. Approaching this field like a "lone eagle," he used his kaleidoscopic mind and his legendary memory, dexterity, and patience to eagerly perform whatever experiments were necessary to come up with his own ideas and theories.

At the same time, while most of his contemporaries were indulging in popularized electrical pontifications of the day, he developed a style of dispassionately questioning them and boldly challenging them.... Possessing this perspective enabled Tom to gradually establish a unique foothold in the world of *practical* science and invention. In fact, at the dawn of the "Age of Light and Power," nothing would serve his destiny any more.

Returning to the story of his youth, by age 12, Tom had seemingly become a virtual adult. He not only talked his parents into letting him go to work selling newspapers, snacks, and candy on the railroad, he had started an entirely separate business selling fruits and vegetables. At age 14 - during the time of the famous pre-Civil War debates between Lincoln and Douglas - he exploited his access to the associated news releases that were being teletyped into the station each day and published them in his own little newspaper called the Weekly Herald.

By focusing upon such "scoops," he ultimately enticed over 300 commuters to subscribe to his splendid little paper Interestingly, because this was the first such publication ever to be type-set, printed, and sold on a train anywhere, an English journal now gave him his first exposure to international notoriety when it related this story in 1860. After his hero, Abraham Lincoln, was finally nominated for president, Tom not only distributed campaign literature on his behalf, he peddled flattering photographs of "the great emancipator." (Interestingly, some 25 years later, Tom's associated feelings about abolition caused him to select Brockton, Massachusetts as the first place to model the first standardized central power system, described elsewhere on this web site.) At its peak, Tom's mini-publishing venture netted him more than ten dollars per day. Because this was considerably more than enough to provide for his own support, he had a good deal of extra income, most of which went towards outfitting the chemical laboratory he had set up in the basement of his home. When his usually tolerant mother finally complained about the odours and danger of all the "poisons" he was amassing, he transferred most of them to a locked room in the basement and put the remainder in his locker room on the train.

One day, while traversing a bumpy section of track, the train lurched, causing a stick of phosphorous to roll onto the floor and ignite. Within moments, the baggage car caught fire. The conductor was so angry, he severely chastised the boy and struck him with a powerful blow on the side of his head. Purportedly, this aggravated the loss of hearing he had experienced earlier from a bout of scarlet fever. In any case, Tom was penalized by being restricted to peddling his newspaper to venues in railroad stations along the track.

Late in his 14th year, Tom contracted scarlet fever. While it has never been ascertained, some biographers have surmised that it was the after effects of this condition - and (or) being struck by the conductor - that destroyed most of his hearing. Whatever the cause - it now became virtually impossible for him to acquire knowledge in a typical educational setting. Amazingly, however, he did never seemed to fret a whole lot over the matter. Naturally inclined towards accepting his fate in life - and promptly adapting to whatever he became convinced was out of his control - he simply committed himself to compensating via *alternative* methods.

Ultimately, Tom finally became totally deaf in his left ear, and approximately 80% deaf in his right ear. He once said that the worst thing about this condition was that he was unable to enjoy the beautiful sounds of singing birds. Indeed, he loved the little creatures so much, he later amassed an aviary of over 5,000 of them. In the meantime, he learned to use the silence associated with deafness to greatly enhance his powers of concentration. In fact, not long after he had acquired the means to have an operation that "would have likely restored his hearing," he flatly refused to act upon the option. His rationale was that he was afraid he "would have difficulty re-learning how to channel his thinking in an ever more noisy world." In any event, Tom's career of producing and selling his newspaper on a train finally came to an abrupt end when he and his press were permanently thrown off the vehicle by an irate railroad supervisor. Shaken and confused by the incident, he continued to frequent the station area.

One day, the stationmaster's young son happened to wander onto the tracks in front of an oncoming boxcar. Tom leaped to action. Luckily - as they tumbled away from its oncoming wheels - they ended up being only slightly injured. One of the most significant events in Tom's life now occurred when - as a reward for his heroism - the boy's grateful father taught him how to master the use of Morse code and the telegraph. In the "age of telegraphy," this was akin to being introduced to learning how to use a state-of-the-art computer. By age 15, Tom had pretty much mastered the basics of this fascinating new career and obtained a job as a replacement for one of the thousands of "brass pounders" (telegraph operators) who had gone off to serve in the Civil War. He now had a golden opportunity to enhance his speed and efficiency in sending and receiving code and performing experiments designed to improve this device.

Once the Civil War ended, to his mother's great dismay, Tom decided - that it was time to "seek his fortune." So, over the next few years, he meandered throughout the Central States, supporting himself as a "tramp operator. At age 16, after working in a variety of telegraph offices, where he performed numerous "moonlight" experiments, he finally came up with his first *authentic* invention. Called an "automatic repeater," it transmitted telegraph signals between unmanned stations, allowing virtually anyone to easily and accurately translate code at their own speed and convenience. Curiously, he never patented the initial version of this idea.

In 1868 - after making a name for himself amongst fellow telegraphers for being a rather flamboyant and quick-witted character who enjoyed playing "mostly harmless" practical jokes - he returned home one day ragged and penniless. Sadly, he found his parents in an even worse predicament.... First, his beloved mother was beginning to show signs of insanity "which was probably aggravated by the strains of an often difficult life." Making matters worse, his rather impulsive father had just quit his job and the local bank was about to foreclose on the family homestead. Tom promptly came to grips with the pathos of this situation and - perhaps for the first time in his life - also resolved to come to grips with a number of his own immature shortcomings.

After a good deal of soul searching, he finally decided that the best thing he could do would be to get right back out on his own and try to make some serious money.... Shortly thereafter, Tom accepted the suggestion of a fellow "lightening slinger" named Billy Adams to come East and apply for a permanent job as a telegrapher with the relatively prestigious Western Union Company in Boston. His willingness to travel over a thousand miles from home was at least partly influenced by the fact that he had been given a free rail ticket by the local street railway company for some repairs he had done for them. The most important factor, however, was the fact that Boston was considered to be "the hub of the scientific, educational, and cultural universe at this time...." Throughout the mid-19th century, New England had many features that were analogous to today's Silicon Valley in California. However, instead of being a haven for the thousands of young "tekkies" - who communicate with each other using computers and internet code of today - it was the home of scores of young telegraphers who anxiously stayed abreast of the emerging age of electricity and the telephone etc. by conversing with via Morse code. During these latter days of the "age of the telegraph," Tom toiled 12 hours a day and six days a week for Western Union. Meanwhile, he continued "moonlighting" on his own projects and, within six months, had applied for and received his very first patent. A beautifully constructed electric vote-recording machine, this first "legitimate" invention he was to come up with turned out to be a disaster.

When he tried to market it to members of the Massachusetts Legislature, they thoroughly denigrated it, claiming "its speed in tallying votes would disrupt the delicate political status-quo." The specific issue was that - during times of stress - political groups regularly relied upon the brief delays that were provided by the process of *manually* counting votes to influence and hopefully change the opinions of their colleagues.... "This is exactly what we do not want" a seasoned politician scolded him, adding that "Your invention would not only destroy the only hope the minority would have in influencing legislation, it would deliver them over - bound hand and foot - to the majority." Although Tom was very much disappointed by this turn of events, he immediately grasped the implications. Even though his remarkable invention allowed each voter to instantly cast his vote from his seat - exactly as it was supposed to do - he realized his idea was so far ahead of its time it was completely devoid of any *immediate* sales appeal. Because of his continuing desperate need for money, Tom now made a critically significant adjustment in his, heretofore, relatively naive outlook on the world of business and marketing.... From now on, he vowed, he would "never waste time inventing things that people would not want to buy." It is important to add here that it was during Tom's 17 month stint in Boston that he was first exposed to lectures at Boston Tech (which was founded in 1861 and became the Mass. Institute of Technology in 1916) and the ideas of several associates on the state-of-the-art of "multiplexing" telegraph signals. This theory and related experimental quests involved the transmission of electrical impulses at different frequencies over telegraph wires, producing horn-like simulations of the human voice and even crude images (the first internet?) via an instrument called the *harmonic telegraph.* Not surprisingly, Alexander Graham Bell, who was also living in Boston at the time, was equally fascinated by this exciting new aspect of communication science. And no wonder, the principles surrounding it ultimately led to the invention of the first *articulating* telephone, the first fax machine, the first microphone, etc. During this epiphany, Edison also became very well acquainted with Benjamin Bredding.

The same age as Bell and Edison, this 21 year old genius would soon provide critically important assistance to Bell in perfecting long distance telephony, the first reciprocating telephone, and the magneto phone. A crack electrician, Bredding, with Watson's assistance, later set up the world's first two-way long distance telephone apparatus for his close friend Alexander Graham Bell, who at the time "knew almost nothing about electricity." [*Copyrighted - never before published - tintype of Bredding and Bell in October of 1876 on the day they successfully communicated across Boston's Charles River in the world's first long distance two-way telephone conversation. i.e., "The world's first practical telephone conversation."*] Bredding had originally worked for the well known promoter, George B. Stearns, who - with Bredding's help - had beaten everyone to the punch when he obtained the first patent for a duplex telegraph line. A device that exploits the fact that electromagnetism and the number and direction of wire windings associated with a connection between telegraph keys can influence the current that flows between them, and greatly facilitate two-way telegraphic communication, it powerfully intrigued Edison. Stearns, finally sold the patent for this highly significant cost-cutting invention to Western Union for $750,000. Bredding (and Edison, of course) wound up getting absolutely nothing from the venture. In the meantime, however, Bredding provided his pal, Tom Edison, with his first detailed introduction and understanding of the state-of-the-art of the harmonograph and the multiplex transmitter. Unlike Edison, Bredding was an extremely modest individual with little taste for aggrandizement and self promotion. The pathetic upshot of all this was that - while the caprice associated with the rough and tumble world of patenting inventions in the mid-19th century ultimately crushed Bridden's innately mild and somewhat naive spirit and his extraordinary potential - it merely spurred the tough-minded Edison on to not only improve the duplex transmitter, but to later patent the world's first quadruplex transmitter.

Deeply in debt and about to be fired by Western Union for "not concentrating on his primary responsibilities and doing too much moonlighting," Edison now borrowed $35.00 from his fellow telegrapher and "night owl" pal, Benjamin Bredding, to purchase a steamship ticket to the "more commercially oriented city of New York." During the third week after arriving in "the big apple" Tom (seen above) was purportedly "on the verge of starving to death." At this precipitous juncture, one of the most amazing coincidences in the annals of technological history now began to unfold. Immediately after having begged a cup of tea from a street vendor, Tom began to meander through some of the offices in New York's financial district. Observing that the manager of one of the local brokerage firms was in a panic, he eventually determined that a critically important stock-ticker in his office had just broken down. Noting that no one in the crowd that had gathered around the defective machine seemed to have a clue on how to fix it, he elbowed his way into the scene and grasped a momentary opportunity to have a go at addressing what was wrong himself.

Luckily, since he had been sleeping in the basement of the building for a few days - and doing quite a bit of snooping around - he already had a pretty good idea of what the device was supposed to do. After spending a few seconds confirming exactly how the stock ticker was intended to work in the first place, Tom reached down and manipulated a loose spring back to where it belonged. To everyone's amazement, except Tom's, the device began to run perfectly. The office manager was so ecstatic, he made an on-the-spot decision to hire Edison to make all such repairs for the busy company for a salary of $300.00 per month. This was not only more than what his pal Benjamin Bredding was making back in Boston but twice the going rate for a top electrician in New York City.

Later in life, Edison recalled that the incident was more euphoric than anything he ever experienced in his life because it made him feel as though he had been "suddenly delivered out of abject poverty and into prosperity." Success at last! It should come as no surprise that, during his free time, Edison soon resumed his habit of "moonlighting" with the telegraph, the quadruplex transmitter, the stock-ticker, etc.

Shortly thereafter, he was absolutely astonished - in fact he nearly fainted - when a corporation paid him $40,000 for all of his rights to the latter device. Convinced that no bank would honour the large cheque he was given for it, which was the first "real" money he had ever received for an invention, young Edison walked around for hours in a stupor, staring at it in amazement. Fearful that someone would steal it, he laid the cash out on his bed and stayed up all night, counting it over and over in disbelief. The next day a wise friend told him to deposit it in a bank forthwith and to just forget about it for a while. A few weeks later, Edison wrote a series of poignant letters back home to his father: "How is mother getting along?... I am now in a position to give you some cash... Write and say how much.... Give mother anything she wants...." Interestingly, it was at this time that he also repaid Bredding the $35.00 he had borrowed earlier.

Over the next three years, Edison's progress in creating successful inventions for industry really took off.... For example, in 1874 - with the money he received from the sale of an electrical engineering firm that held several of his patents - he opened his first complete testing and development laboratory in New York, New Jersey. At age 29, he commenced work on the carbon transmitter, which ultimately made Alexander Graham Bell's amazing new "articulating" telephone (which by today's standards sounded more like someone trying to talk through a kazoo than a telephone) audible enough for practical use. Interestingly, at one point during this intense period, Edison was as close to inventing the telephone as Bell was to inventing the phonograph. Nevertheless, shortly after Edison moved his laboratory to Menlo Park, N.J. in 1876, he invented - in 1877 - the first phonograph. In 1879, extremely disappointed by the fact that Bell had beaten him in the race to patent the first authentic transmission of the human voice, Edison now "one upped" all of his competition by inventing the first commercially practical incandescent electric light bulb.

And if that wasn't enough to forever seal his unequaled importance in technological history, he came up with an invention that - in terms of its collective effect upon mankind - has had more impact than any other. In 1883 and 1884, while beating a path from his research lab to the patent office, he introduced the world's first economically viable system of centrally generating and distributing electric light, heat, and power. Powerfully, instrumental in shaping the world we know today, even his harshest critics grant that it was a Herculean achievement that only he was capable of bringing about at this specific point in history. By 1887, Edison was recognized for having set up the world's first full fledged research and development center in West Orange, New Jersey. An amazing enterprise, its significance is as much misunderstood as his work in developing the first practical centralized power system. Regardless, within a year, this fantastic operation was the largest scientific testing laboratory in the world. In 1890, Edison immersed himself in developing the first Vitascope, which would lead to the first silent motion pictures. And, by 1892, his Edison General Electric Co. had fully merged with another firm to become the great General Electric Corporation, in which he was a major stockholder.

At the turn-of-the-century, Edison invented the first practical dictaphone, mimeograph, and storage battery. After creating the "kinetiscope" and the first silent film in 1904, he went on to introduce The Great Train Robbery in 1903, which was a ten minute clip that was his first attempt to blend audio with silent moving images to produce "talking pictures." By now, Edison was being hailed world-wide as The wizard of Menlo Park, The father of the electrical age," and The greatest inventor who ever lived." Naturally, when World War I began, he was asked by the U. S. Government to focus his genius upon creating defensive devices for submarines and ships. During this time, he also perfected a number of important inventions relating to the enhanced use of rubber, concrete, and ethanol. By the 1920s Edison was internationally revered. However, even though he was personally acquainted with scores of very important people of his era, he cultivated very few close friendships. And due to the continuing demands of his career, there were still relatively long periods when he spent a shockingly small amount of time with his family. It wasn't until his health began to fail, in the late 1920s, that Edison finally began to slow down and, so to speak, "smell the flowers." Up until obtaining his last (1,093rd) patent at age 83, he worked mostly at home where, though increasingly frail, he enjoyed greeting former associates and famous people such as Charles Lindberg, Marie Curie, Henry Ford, and President Herbert Hoover etc. He also enjoyed reading the mail of admirers and puttering around when he was able in his office and home laboratory. Thomas Edison died At 9 pm. on October 18th, 1931 in New Jersey. He was 84 years of age. Shortly before passing away, he awoke from a coma and quietly whispered to his wife, Mina, who had been keeping a vigil all night by his side: "It is very beautiful over there..." Recognizing that his death marked the end of an era in the progress of civilization, countless individuals, communities, and corporations throughout the world dimmed their lights and, or, briefly turned off their electric power in his honour on the evening of the day he was laid to rest at his beautiful estate at Glenmont, New Jersey. Most realized that even though he was far from being a flawless human being, and may not have really had the avuncular personality that was so often proscribed to him by myth makers, he was an essentially good man with a powerful mission.... Utterly driven by a superhuman desire to fulfill the promise of research and invent things to serve mankind, no one did more to help realize our Puritan founders dream of creating a country that - at its best - was viewed by the rest of the world as "a shining city upon a hill."

JOHN LOGIE BAIRD

John Logie Baird was born on August 13th, 1888, in Helensburgh, Dunbarton, Scotland and died on June 14th, 1946, in Bexhill-on-Sea, Sussex, England. John Logie Baird received a diploma course in electrical engineering at the Glasgow and West of Scotland Technical College (now called Strathclyde University), and studied towards his Bachelor of Science Degree in electrical engineering from the University of Glasgow, interrupted by the outbreak of W.W.I.

John Logie Baird is remembered as being an inventor of a mechanical television system. In the 1920's, John Logie Baird and American Clarence W. Hansell patented the idea of using arrays of transparent rods to transmit images for television and facsimiles respectively. Baird's 30 line images were the first demonstrations of television by reflected light rather than back-lit silhouettes. John Logie Baird based his technology on Paul Nipkow's scanning disc idea and later developments in electronics.

The television pioneer created the first televised pictures of objects in motion (1924), the first televised human face (1925) and a year later he televised the first moving object image at the Royal Institution in London. His 1928 transatlantic transmission of the image of a human face was a broadcasting milestone. Colour television (1928), stereoscopic television and television by infra-red light were all demonstrated by Baird before 1930. He successfully lobbied for broadcast time with the British Broadcasting Company and the BBC started broadcasting television on the Baird 30-line system in 1929. The first simultaneous sound and vision telecast was broadcast in 1930. In July 1930, the first British Television Play was transmitted, "The Man with the Flower in his Mouth."

In 1936, the British Broadcasting Corporation adopted television service using the electronic television technology of

Marconi-EMI (the world's first regular high-resolution service - 405 lines per picture), it was that technology that won out over Baird's system.

John Logie Baird, a Scottish engineer and entrepreneur, achieved his first transmissions of simple face shapes in 1924 using mechanical television. On March 25, 1925, Baird held his first public demonstration of television at the London department store Selfridges on Oxford Street in London. In this demonstration, he had not yet obtained adequate half tones in the moving pictures, and only silhouettes were visible.

In the first week of October 1925, Baird obtained the first actual television picture in his laboratory. At this time, his test subject was a ventriloquist's dummy, Stooky Bill, which was placed in front of the camera apparatus. Baird later recollected, "The image of the dummy's head formed itself on the screen with what appeared to me an almost unbelievable clarity. I had got it! I could scarcely believe my eyes and felt myself shaking with excitement."

After much discussion with his business associates, and further improvements, Baird decided to publicly demonstrate television on Tuesday 26th January 1926, again at Selfridges' department store. This was the first opportunity for the general public to see television. The Baird Company continued to publicise this historic demonstration, and J.L. Baird's other scientific breakthroughs as they feverishly worked to obtain financial backing and construct a line of home receivers.

With Baird's transmitting equipment, the British Broadcasting Corporation began regular experimental television broadcasts on September 30, 1929. By the following year, most of Britain's major radio dealers were selling Baird kits and ready-made receivers through retail and by mail order.

GEORGE EASTMAN

George Eastman (1854-1932), American inventor and philanthropist, who played a leading role in transforming photography from an expensive hobby of a few devotees into a relatively inexpensive and immensely popular pastime. He was born in Waterville, New York, and was self-educated. In 1884 Eastman patented the first film in roll form to prove practicable; in 1888 he perfected the Kodak camera, the first camera designed specifically for roll film. In 1892 he established the Eastman Kodak Company, at Rochester, New York, one of the first firms to mass-produce standardized photography equipment. This company also manufactured the flexible transparent film, devised by Eastman in 1889, which proved vital to the subsequent development of the motion picture industry.

Eastman was associated with the company in an administrative and an executive capacity until his death and contributed much to the development of its notable research facilities. He was also one of the outstanding philanthropists of his time, donating more than $75 million to various projects. Notable among his contributions were a gift to the Massachusetts Institute of Technology and endowments for the establishment of the Eastman School of Music in 1918 and a school of medicine and dentistry in 1921 at the University of Rochester.

George Eastman invented dry, transparent, and flexible, photographic film (rolled photography film) and the Kodak cameras that could use the new film in 1888. George Eastman, an avid photographer was the founder of the Kodak company.

"You press the button, we do the rest" promised George Eastman in 1888 with this advertising slogan for his Kodak camera.

Eastman wanted to simplify photography and make it available to everyone. In 1883, Eastman announced film in rolls.

"Kodak" was born in 1888 when the first Kodak camera entered the market. Pre-loaded with enough film for 100 exposures, the camera could easily be carried and handheld for operation. After exposure, the whole camera was returned to the company in Rochester, New York, where the film was developed, prints were made, new film was inserted, and then returned to the customer.

"The letter *K* had been a favourite with me—it seems a strong, incisive sort of letter. It became a question of trying out a great number of combinations of letters that made words starting and ending with *K*." -George Eastman

Eastman was one of the first American industrialists to employ a full-time research scientist. Together with his associate, Eastman perfected the first commercial transparent roll film which made possible Thomas Edison's motion picture camera in 1891.

Patent Suits

On April 26 1976, one of the largest patent suits involving photography was filed in the U.S. District Court of Massachusetts. Polaroid Corporation, the assignee of numerous patents relating to instant photography, brought an action against Kodak Corporation for infringement of 12 Polaroid patents relating to instant photography. On October 11 1985, after five years of vigorous pretrial activity and 75 days of trial, seven Polaroid patents were found to be valid and infringed. Kodak was out of the instant picture market leaving customers with useless cameras and no film. Kodak offered camera owners various compensation for their loss.

JOHN PEMBERTON

John Pemberton was the son of James Clifford Pemberton and Martha L. Worsham Gent in Knoxville. The Pembertons married on July 20, 1828, in Crawford County, and John was born on July 8, 1831. The Pemberton family moved to Rome, and John attended medical school in Macon, receiving his degree at the age of 19. Sometime later he received a graduate degree in pharmacy. He married Wesleyan student Ann Eliza Clifford Lewis and moved to Columbus in 1853. The couple had one son, Charles, born in 1854.

John was a druggist in Columbus and built a laboratory where he made and sold medicines, photographic chemicals, and cosmetic products including a popular perfume he called Sweet Southern Bouquet. He moved his family to Atlanta in 1870.

John's mother Martha was the daughter of Archer Worsham and Nancy Clark Smith, who lived in Baldwin Co. Her brother Archibald and her sister Virginia Ann were married to siblings of Crawford Countian Richard Waller Ellis, whose family also came from Baldwin. James was the son of John Pemberton and Rebecca Clifford. Their other children were Ann, Archy, Mary, and Martha. Pemberton served on the first pharmacy licensing board in the state, established a modern chemical laboratory that was the first state-run facility to test soil and crop chemicals, and was a trustee of Emory University School of Medicine. He fought for the Confederacy, rising to the rank of lieutenant colonel in the 23rd Georgia Cavalry Battallion. Some sources say his severe wounds from the war led to morphine and cocaine addictions.

John began work on a coca and cola nut-based nerve tonic called Pemberton's French Wine Cola when he was a druggist and chemist in Columbus. In 1866 he was selling the product through Atlanta druggists when the city passed a prohibition law. So he adjusted the formula, renamed the product Coca-Cola (as suggested by partner Frank Robinson), and marketed it as both a "delicious, exhilarating, refreshing and invigorating" soda-fountain and a "temperance drink."

When the prohibition law was repealed after just one year, Pemberton returned to concentration on his medicinal coca and wine formula, leaving development of the new fountain drink to his son Charles. He died on Aug. 16, 1888, only a few months after the Coca-Cola Co. was incorporated. He is buried in Linwood Cemetery in Columbus.

John Pemberton concocted the Coca Cola formula in a three legged brass kettle in his backyard. The name was a suggestion given by John Pemberton's bookkeeper Frank Robinson. Being a bookkeeper, Frank Robinson also had excellent penmanship. It was he who first scripted "Coca Cola" into the flowing letters which has become the famous logo of today.

The soft drink was first sold to the public at the soda fountain in Jacob's Pharmacy in Atlanta on May 8, 1886.

About nine servings of the soft drink were sold each day. Sales for that first year added up to a total of about $50. The funny thing was that it cost John Pemberton over $70 in expenses, so the first year of sales were a loss.

Until 1905, the soft drink, marketed as a tonic, contained extracts of cocaine as well as the caffeine-rich kola nut.

By the late 1890s, Coca-Cola was one of America's most popular fountain drinks. With another Atlanta pharmacist, Asa Griggs Candler, at the helm, the Coca-Cola Company increased syrup sales by over 4000% between 1890 and 1900. Advertising, was an important factor in Pemberton and Candler's success and by the turn of the century, the drink was sold across the United States and Canada. Around the same time, the company began selling syrup to independent bottling companies licensed to sell the drink. Even today, the US soft drink industry is organized on this principle.

Until the 1960s, both small town and big city dwellers enjoyed carbonated beverages at the local soda fountain or ice cream saloon. Often housed in the drug store, the soda fountain counter served as a meeting place for people of all ages. Often combined with lunch counters, the soda fountain declined in popularity as commercial ice cream, bottled soft drinks, and fast food restaurants came to the fore.

On April 23, 1985, the trade secret "New Coke" formula was released. Today, products of the Coca Cola Company are consumed at the rate of more than one billion drinks per day.

HENRY FORD

Henry Ford, born July 30, 1863, was the first of William and Mary Ford's six children. He grew up on a prosperous family farm in what is today Dearborn, Michigan. Henry enjoyed a childhood typical of the rural nineteenth century, spending days in a one-room school and doing farm chores. At an early age, he showed an interest in mechanical things and a dislike for farm work.

In 1879, sixteen-year-old Ford left home for the nearby city of Detroit to work as an apprentice machinist, although he did occasionally return to help on the farm. He remained an apprentice for three years and then returned to Dearborn. During the next few years, Henry divided his time between operating or repairing steam engines, finding occasional work in a Detroit factory, and over-hauling his father's farm implements, as well as lending a reluctant hand with other farm work. Upon his marriage to Clara Bryant in 1888, Henry supported himself and his wife by running a sawmill.

In 1891, Ford became an engineer with the Edison Illuminating Company in Detroit. This event signified a conscious decision on Ford's part to dedicate his life to industrial pursuits. His promotion to Chief Engineer in 1893 gave him enough time and money to devote attention to his personal experiments on internal combustion engines.

These experiments culminated in 1896 with the completion of his own self-propelled vehicle–the Quadricycle. The Quadricycle had four wire wheels that looked like heavy bicycle wheels, was steered with a tiller like a boat, and had only two forward speeds with no reverse. Although Ford was not the first to build a self-propelled vehicle with a gasoline engine, he was, however, one of several automotive pioneers who helped this country become a nation of motorists.

After two unsuccessful attempts to establish a company to manufacture automobiles, the Ford Motor Company was incorporated in 1903 with Henry Ford as vice-president and chief engineer. The infant company produced only a few cars a day at the Ford factory on Mack Avenue in Detroit. Groups of two or three men worked on each car from components made to order by other companies.

Henry Ford realized his dream of producing an automobile that was reasonably priced, reliable, and efficient with the introduction of the Model 'T' in 1908. This vehicle initiated a new era in personal transportation. It was easy to operate, maintain, and handle on rough roads, immediately becoming a huge success.

By 1918, half of all cars in America were Model T's. To meet the growing demand for the Model 'T', the company opened a large factory at Highland Park, Michigan, in 1910. Here, Henry Ford combined precision manufacturing, standardized and interchangeable parts, a division of labour, and, in 1913, a continuous moving assembly line. Workers remained in place, adding one component to each automobile as it moved past them on the line. Delivery of parts by conveyor belt to the workers was carefully timed to keep the assembly line moving smoothly and efficiently. The introduction of the moving assembly line revolutionized automobile production by significantly reducing assembly time per vehicle, thus lowering costs. Ford's production of Model T's made his company the largest automobile manufacturer in the world.

The company began construction of the world's largest industrial complex along the banks of the Rouge River in Dearborn, Michigan, during the late 1910s and early 1920s. The massive Rouge Plant included all the elements needed for automobile production: a steel mill, glass factory, and automobile assembly line. Iron ore and coal were brought in on Great Lakes steamers and by railroad, and were used to produce both iron and steel. Rolling mills, forges, and assembly shops transformed the steel into springs, axles, and car bodies.

Foundries converted iron into engine blocks and cylinder heads that were assembled with other components into engines. By September 1927, all steps in the manufacturing process from refining raw materials to final assembly of the automobile took place at the vast Rouge Plant, characterising Henry Ford's idea of mass production.

JESSE WILFORD RENO

Jesse Wilford Reno, born in 1861 in Fort Leavenworth, Kansas, was an inventive young man who formulated his idea for an inclined moving stairway at age 16. After graduating from Lehigh University in Pennsylvania, his engineering career took him to Colorado, then to Americus, Georgia where he is credited with building the first electric railway in the southern U.S.

Reno submitted his first patent application for a "new and useful endless conveyor or elevator" in 1891. It became effective 15 months later. The machine was built and installed at Coney Island, Brooklyn, as an amusement ride in September 1895. Moving stairways were just one arrow in the quiver, for in 1896, Reno developed plans for the building of the New York City subway, a double-decker underground system that could be completed in three years. With the plan not accepted, the inventor married and moved to London where he opened his new company, The Reno Electric Stairways and Conveyors, Ltd. in 1902.

His pallet-type moving stairways were being installed throughout the U.S., Great Britain and Europe, but Reno became fascinated with a new challenge — building the first Spiral Moving Walkway. He joined with William Henry Aston, holder of a patent for the flexible pallet coupling and chain, to create the pioneering mechanism that was exhibited for four years and installed on the London railway at his own cost, but never used by the public. In 1903, the firm of Waygood and Otis Limited bought a third share in the Reno Company, but with the failure of the Spiral Walkway, Reno sold his patents to Otis and returned to the U.S.

GEORGE WASHINGTON CARVER

From inauspicious and dramatic beginnings, George Washington Carver became one of the nation's greatest educators and agricultural researchers. He was born in about 1864 (the exact year is unknown) on the Moses Carver plantation in Diamond Grove. His father died in an accident shortly before his birth, and when he was still an infant, Carver and his mother were kidnapped by slave raiders. The baby was returned to the plantation, but his mother was never heard from again.

Carver grew to be a student of life and a scholar, despite the illness and frailty of his early childhood. Because he was not strong enough to work in the fields, he helped with household chores and gardening. Probably as a result of these duties and because of the hours he would spend exploring the woods around his home, he developed a keen interest in plants at an early age. He gathered and cared for a wide variety of flora from the land near his home and became known as the 'plant doctor', helping neighbours and friends with ailing plants. He learned to read, write and spell at home because there were no schools for African Americans in Diamond Grove. From age 10, his thirst for knowledge and desire for formal education led him to several communities in Missouri and Kansas and finally, in 1890, to Indiana, Iowa, were he enrolled at Simpson College to study piano and painting.

He excelled in art and music, but art instructor Etta Budd, whose father was head of the Iowa State College Department of Horticulture, recognized Carver's horticultural talents. She convinced him to pursue a more pragmatic career in scientific agriculture and, in 1891, he became the first African American to enroll at Iowa State College of Agriculture and Mechanic Arts, which today is Iowa State University.

Through quiet determination and perseverance, Carver soon became involved in all facets of campus life. He was a leader in the YMCA and the debate club. He worked in the dining rooms and as a trainer for the athletic teams. He was captain, the highest student rank, of the campus military regiment. His poetry was published in the student newspaper and two of his paintings were exhibited at the 1893 World's Fair in Chicago.

Carver's interests in music and art remained strong, but it was his excellence in botany and horticulture that prompted professors Joseph Budd and Louis Pammel to encourage him to stay on as a graduate student after he completed his bachelor's degree in 1894. Because of his proficiency in plant breeding, Carver was appointed to the faculty, becoming Iowa State's first African American faculty member.

Over the next two years, as assistant botanist for the College Experiment Station, Carver quickly developed scientific skills in plant pathology and mycology, the branch of botany that deals with fungi. He published several articles on his work and gained national respect. In 1896, he completed his master's degree and was invited by Booker T. Washington to join the faculty of Alabama's Tuskegee Institute.

At Tuskegee, he gained an international reputation in research, teaching and outreach. Carver taught his students that nature is the greatest teacher and that by understanding the forces in nature, one can understand the dynamics of agriculture. He instilled in them the attitude of gentleness and taught that education should be "made common" —used for betterment of the people in the community.

Carver's work resulted in the creation of 325 products from peanuts, more than 100 products from sweet potatoes and hundreds more from a dozen other plants native to the South. These products contributed to rural economic improvement by offering alternative crops to cotton that were beneficial for the farmers and for the land. During this time, Carver also carried the Iowa State extension concept to the South and created "movable schools," bringing practical agricultural knowledge to farmers, thereby promoting health, sound nutrition and self-sufficiency. Dennis Keeney, director of the Leopold Center for Sustainable Agriculture at Iowa State University, writes in the *Leopold Letter* newsletter about Carver's contributions:

> *Carver worked on improving soils, growing crops with low inputs, and using species that fixed nitrogen (hence, the work on the cowpea and the peanut). Carver wrote in* The Need of Scientific Agriculture in the South*: "The virgin fertility of our soils and the vast amount of unskilled labour have been more of a curse than a blessing to agriculture. This exhaustive system for cultivation, the destruction of forest, the rapid and almost constant decomposition of organic matter, have made our agricultural problem one requiring more brains than of the North, East or West."*

Carver died in 1943. He received many honours in his lifetime and after, including a 1938 feature film, *Life of George Washington Carver*; the George Washington Carver Museum, dedicated at Tuskegee Institute in 1941; the Roosevelt Medal for Outstanding Contribution to Southern Agriculture in 1942; a national monument in Diamond Grove, Mo., commemorative postage stamps in 1947 and 1998; and a fifty-cent coin in 1951. He was elected to the Hall of Fame for Great Americans in 1977 and inducted into the National Inventors Hall of Fame in 1990. In 1994, Iowa State awarded him the degree, Doctor of Humane Letters.

WILBUR & ORVILLE WRIGHT

In 1899, after Wilbur Wright had written a letter of request to the Smithsonian Institution for information about flight experiments, the Wright Brothers designed their first aircraft: a small, biplane glider flown as a kite to test their solution for controlling the craft by wing warping. Wing warping is a method of arching the wingtips slightly to control the aircraft's rolling motion and balance.

The Wrights spent a great deal of time observing birds in flight. They noticed that birds soared into the wind and that the air flowing over the curved surface of their wings created lift. Birds change the shape of their wings to turn and maneuver. They believed that they could use this technique to obtain roll control by warping, or changing the shape, of a portion of the wing.

Over the next three years, Wilbur and his brother Orville would design a series of gliders which would be flown in both unmanned (as kites) and piloted flights. They read about the works of Cayley, and Langley, and the hang-gliding flights of Otto Lilienthal. They corresponded with Octave Chanute concerning some of their ideas. They recognized that control of the flying aircraft would be the most crucial and hardest problem to solve.

Following a successful glider test, the Wrights built and tested a full-size glider. They selected Kitty Hawk, North Carolina as their test site because of its wind, sand, hilly terrain and remote location.

In 1900, the Wrights successfully tested their new 50-pound biplane glider with its 17-foot wingspan and wing-warping mechanism at Kitty Hawk, in both unmanned and piloted flights. In fact, it was the first piloted glider. Based upon the results, the

Wright Brothers planned to refine the controls and landing gear, and build a bigger glider.

In 1901, at Kill Devil Hills, North Carolina, the Wright Brothers flew the largest glider ever flown, with a 22-foot wingspan, a weight of nearly 100 pounds and skids for landing. However, many problems occurred: the wings did not have enough lifting power; forward elevator was not effective in controlling the pitch; and the wing-warping mechanism occasionally caused the airplane to spin out of control. In their disappointment, they predicted that man will probably not fly in their lifetime.

In spite of the problems with their last attempts at flight, the Wrights reviewed their test results and determined that the calculations they had used were not reliable. They decided to build a wind tunnel to test a variety of wing shapes and their effect on lift. Based upon these tests, the inventors had a greater understanding of how an airfoil (wing) works and could calculate with greater accuracy how well a particular wing design would fly. They planned to design a new glider with a 32-foot wingspan and a tail to help stabilize it.

During 1902, the brothers flew numerous test glides using their new glider. Their studies showed that a movable tail would help balance the craft and the Wright Brothers connected a movable tail to the wing-warping wires to coordinate turns. With successful glides to verify their wind tunnel tests, the inventors planned to build a powered aircraft.

After months of studying how propellers work the Wright Brothers designed a motor and a new aircraft sturdy enough to accommodate the motor's weight and vibrations. The craft weighed 700 pounds and came to be known as the Flyer.

The brothers built a movable track to help launch the Flyer. This downhill track would help the aircraft gain enough airspeed to fly. After two attempts to fly this machine, one of which resulted in a minor crash, Orville Wright took the Flyer for a 12-second, sustained flight on December 17, 1903. This was the first successful, powered, piloted flight in history.

In 1904, the first flight lasting more than five minutes took place on November 9. The Flyer II was flown by Wilbur Wright.

In 1908, passenger flight took a turn for the worse when the first fatal air crash occurred on September 17. Orville Wright was piloting the plane. Orville Wright survived the crash, but his passenger, Signal Corps Lieutenant Thomas Selfridge, did not. The Wright Brothers had been allowing passengers to fly with them since May 14, 1908.

In 1909, the U.S. Government bought its first airplane, a Wright Brothers biplane, on July 30. The airplane sold for $25,000 plus a bonus of $5,000 because it exceeded 40 mph.

In 1911, the Wrights' Vin Fiz was the first airplane to cross the United States. The flight took 84 days, stopping 70 times. It crash-landed so many times that little of its original building materials were still on the plane when it arrived in California. The Vin Fiz was named after a grape soda made by the Armour Packing Company.

In 1912, a Wright Brothers plane, the first airplane armed with a machine gun was flown at an airport in College Park, Maryland. The airport had existed since 1909 when the Wright Brothers took their government-purchased airplane there to teach Army officers to fly.

On July 18, 1914, an Aviation Section of the Signal Corps (part of the Army) was established. Its flying unit contained airplanes made by the Wright Brothers as well as some made by their chief competitor, Glenn Curtiss.

That same year, the U.S. Court has decided in favour of the Wright Brothers in a patent suit against Glenn Curtiss. The issue concerned lateral control of aircraft, for which the Wrights maintained they held patents.

Although Curtiss's invention, ailerons (French for "little wing"), was far different from the Wrights' wing-warping mechanism, the Court determined that use of lateral controls by others was 'unauthorized' by patent law.

ENRICO FERMI

Enrico Fermi was born in Rome on 29th September, 1901, the son of Alberto Fermi, a Chief Inspector of the Ministry of Communications, and Ida de Gattis. He attended a local grammar school, and his early aptitude for mathematics and physics was recognized and encouraged by his father's colleagues, among them A. Amidei. In 1918, he won a fellowship of the Scuola Normale Superiore of Pisa. He spent four years at the University of Pisa, gaining his doctor's degree in physics in 1922, with Professor Puccianti.

Soon afterwards, in 1923, he was awarded a scholarship from the Italian Government and spent some months with Professor Max Born in Gottingen. With a Rockefeller Fellowship, in 1924, he moved to Leyden to work with P. Ehrenfest, and later that same year he returned to Italy to occupy for two years (1924-1926) the post of Lecturer in Mathematical Physics and Mechanics at the University of Florence.

In 1926, Fermi discovered the statistical laws, nowadays known as the Fermi statistics, governing the particles subject to Pauli's exclusion principle (now referred to as fermions, in contrast with bosons which obey the Bose-Einstein statistics).

In 1927, Fermi was elected Professor of Theoretical Physics at the University of Rome (a post which he retained until 1938, when he - immediately after the receipt of the Nobel Prize - emigrated to America, primarily to escape Mussolini's fascist dictatorship).

During the early years of his career in Rome he occupied himself with electrodynamic problems and with theoretical investigations on various spectroscopic phenomena. But a capital turning-point came when he directed his attention from the outer electrons towards the atomic nucleus itself. In 1934, he evolved the β-decay theory, coalescing previous work on radiation theory with Pauli's idea of the neutrino. Following the discovery by Curie and Joliot of artificial radioactivity (1934), he demonstrated that nuclear transformation occurs in almost every element subjected to neutron bombardment. This work resulted in the discovery of slow neutrons that same year, leading to the discovery of nuclear fission and the production of elements lying beyond what was until then the Periodic Table.

In 1938, Fermi was without doubt the greatest expert on neutrons, and he continued his work on this topic on his arrival in the United States, where he was soon appointed Professor of Physics at Columbia University, New York (1939-1942).

Upon the discovery of fission, by Hahn and Strassmann early in 1939, he immediately saw the possibility of emission of secondary neutrons and of a chain reaction. He proceeded to work with tremendous enthusiasm, and directed a classical series of experiments which ultimately led to the atomic pile and the first controlled nuclear chain reaction. This took place in Chicago on December 2, 1942 - on a squash court situated beneath Chicago's stadium. He subsequently played an important part in solving the problems connected with the development of the first atomic bomb (He was one of the leaders of the team of physicists on the Manhattan Project for the development of nuclear energy and the atomic bomb.)

In 1944, Fermi became American citizen, and at the end of the war (1946) he accepted a professorship at the Institute for Nuclear Studies of the University of Chicago, a position which he held until his untimely death in 1954. There he turned his attention to high-energy physics, and led investigations into the pion-nucleon interaction.

During the last years of his life Fermi occupied himself with the problem of the mysterious origin of cosmic rays, thereby developing a theory, according to which a universal magnetic field - acting as a giant accelerator - would account for the fantastic energies present in the cosmic ray particles.

Professor Fermi was the author of numerous papers both in theoretical and experimental physics. His most important contributions were:

"Sulla quantizzazione del gas perfetto monoatomico", *Rend. Accad. Naz. Lincei,* 1935 (also in *Z. Phys.,* 1936), concerning the foundations of the statistics of the electronic gas and of the gases made of particles that obey the Pauli Principle.

Several papers published in *Rend. Accad. Naz. Lincei,* 1927-28, deal with the statistical model of the atom (Thomas-Fermi atom model) and give a semiquantitative method for the calculation of atomic properties. A résumé of this work was published by Fermi in the volume: *Quantentheorie und Chemie,* edited by H. Falkenhagen, Leipzig, 1928.

"Uber die magnetischen Momente der AtomKerne", *Z. Phys.,* 1930, is a quantitative theory of the hyperfine structures of spectrum lines. The magnetic moments of some nuclei are deduced therefrom.

"Tentativo di una teoria dei raggi β", *Ricerca Scientifica,* 1933 (also *Z. Phys.,* 1934) proposes a theory of the emission of β-rays, based on the hypothesis, first proposed by Pauli, of the existence of the neutrino.

The Nobel Prize for Physics was awarded to Fermi for his work on the artificial radioactivity produced by neutrons, and for nuclear reactions brought about by slow neutrons. The first paper on this subject *"Radioattivita indotta dal bombardamento di neutroni"* was published by him in *Ricerca Scientifica,* 1934. All the work is collected in the following papers by himself and various collaborators: "Artificial radioactivity produced by neutron bombardment", *Proc. Roy. Soc.,* 1934 and 1935; "On the absorption and diffusion of slow neutrons", *Phys. Rev.,* 1936. The theoretical problems connected with the neutron are discussed by Fermi in the paper *"Sul moto dei neutroni lenti", Ricerca Scientifica,* 1936.

His *Collected Papers* are being published by a Committee under the Chairmanship of his friend and former pupil, Professor E. Segre (Nobel Prize winner 1959, with O. Chamberlain, for the discovery of the antiproton).

Fermi was member of several academies and learned societies in Italy and abroad (he was early in his career, in 1929, chosen among the first 30 members of the Royal Academy of Italy).

As lecturer he was always in great demand (he has also given several courses at the University of Michigan, Ann Arbor; and Stanford University, Calif.). He was the first recipient of a special award of $50,000 - which now bears his name - for work on the atom.

Professor Fermi married Laura Capon in 1928. They had one son Giulio and one daughter Nella. His favourite pastimes were walking, mountaineering, and winter sports.

He died in Chicago on 29th November, 1954.

STEPHEN WOZNIAK

"The first Apple was just a culmination of my whole life." - said Stephen Wozniak, Co-Founder, Apple Computers.

Following the introduction of the Altair, a boom in personal computers occurred, and luckily for the consumer, the next round of home computers were considered useful and a joy to use.

In 1975, Steve Wozniak was working for Hewlett Packard (calculator manufacturers) by day and playing computer hobbyist by night, tinkering with the early computer kits like the Altair. "All the little computer kits that were being touted to hobbyists in 1975 were square or rectangular boxes with non understandable switches on them..." claimed Wozniak. Wozniak realized that the prices of some computer parts (e.g. microprocessors and memory chips) had got so low that he could buy them with maybe a month's salary. Wozniak decided that, with some help from fellow hobbyist Steve Jobs, they could build their own computer.

On April Fool's Day, 1976, Steve Wozniak and Steve Jobs released the Apple I computer and started Apple Computers. The Apple I was the first single circuit board computer. It came with a video interface, 8kb of RAM and a keyboard. The system incorporated some economical components, including the 6502 processor (only $25 dollars - designed by Rockwell and produced by MOS Technologies) and dynamic RAM.

The pair showed the prototype Apple I, mounted on plywood with all the components visible, at a meeting of a local computer hobbyist group called "The Homebrew Computer Club" (based in Palo Alto, California). A local computer dealer (The Byte Shop) saw it and ordered 100 units, providing that Wozniak and Jobs agreed to assemble the kits for the customers. About two hundred Apple I computers were built and sold over a ten month period, for the superstitious price of $666.66.

In 1977, Apple Computers was incorporated and the Apple II computer model was released. The first West Coast Computer Faire was held in San Francisco the same year, and attendees saw the public debut of the Apple II (available for $1298). The Apple II was also based on the 6502 processor, but it had colour graphics (a first for a personal computer), and used an audiocassette drive for storage. Its original configuration came with 4 kb of RAM, but a year later this was increased to 48 kb of RAM and the cassette drive was replaced by a floppy disk drive.

The Commodore PET

The Commodore PET (Personal Electronic Transactor or maybe rumoured to be named after the "pet rock" fad) was designed by Chuck Peddle. It was first presented at the January, 1977, Winter Consumer Electronics Show and later at the West Coast Computer Faire. The Pet Computer also ran on the 6502 chip, but it cost only $795, half the price of the Apple II. It included 4 kb of RAM, monochrome graphics and an audio cassette drive for data storage. Included was a version of BASIC in 14kb of ROM. Microsoft developed its first 6502-based BASIC for the PET and then sold the source code to Apple for AppleBASIC. The keyboard, cassette drive and small monochrome display all fit within the same self contained unit.

Note: Steve Jobs and Steve Wozniak at one point in time showed the Apple I prototype to Commodore, who agreed to buy Apple. Steve Jobs then decided not to sell to Commodore, who bought MOS Technology instead and then designed the PET. The Commodore PET was seen at the time to be a chief rival of the Apple.

In 1977, Radio Shack introduced its TRS-80 microcomputer, also nicknamed the "Trash-80". It was based on the Zilog Z80 processor (an 8-bit microprocessor whose instruction set is a superset of the Intel 8080) and came with 4 kb of RAM and 4 kb of ROM with BASIC. An optional expansion box enabled memory expansion, and audio cassettes were used for data storage, similar to the PET and the first Apples. Over 10,000 TRS-80s were sold during the first month of production. The later TRS-80 Model II came complete with a disk drive for program and data storage. At that time, only Apple and Radio Shack had machines with disk drives. With the introduction of the disk drive, applications for the personal computer proliferated as distribution of software became easier.

Born in 1955 Los Altos CA; Evangelic bad boy who, with Steve Wozniak, co-founded Apple Computer Corporation and became a multimillionaire before the age of 30. Subsequently started the NeXT Corporation to provide an educational system at a reasonable price, but found that software was a better seller than hardware.

Going to work for Atari after leaving Reed College, Jobs renewed his friendship with Steve Wozniak. The two designed computer games for Atari and a telephone "blue box", getting much of their impetus from the Homebrew Computer Club. Beginning work in the Job's family garage they managed to make their first "killing" when the Byte Shop in Mountain View bought their first fifty fully assembled computers. On this basis the Apple Corporation was founded, the name based on Job's favourite fruit and the logo (initially used as the unregistered logo of the ACM APL Conference in San Francisco) chosen to play on both the company name and the word byte. Through the early 1980's Jobs controlled the business side of the corporation, successively hiring presidents who would take the organization to a higher level. With the layoffs of 1985 Jobs lost a power struggle with John Sculley, and after a short hiatus reappeared with new funding to create the NeXT corporation.

STEVE JOBS

Steve Jobs innovative idea of a personal computer led him into revolutionizing the computer hardware and software industry. When Jobs was twenty one, he and a friend, Wozniak, built a personal computer called the Apple. The Apple changed people's idea of a computer from a gigantic and inscrutable mass of vacuum tubes only used by big business and the government to a small box used by ordinary people. No company has done more to democratize the computer and make it user-friendly than Apple Computer Inc. Jobs software development for the Macintosh re-introduced windows interface and mouse technology which set a standard for all applications interface in software.

Two years after building the Apple I, Jobs introduced the Apple II. The Apple II was the best buy in personal computers for home and small business throughout the following five years. When the Macintosh was introduced in 1984, it was marketed towards medium and large businesses. The Macintosh took the first major step in adapting the personal computer to the needs of the corporate work force. Workers lacking computer knowledge accomplished daily office activities through the Macintosh's user-friendly windows interface. Steve Jobs was considered a brilliant young man in Silicon Valley, because he saw the future demands of the computer industry. He was able to build a personal computer and market the product. "The personal computer was created by the hardware revolution of the 1970's and the next dramatic change will come from a software revolution," said Jobs. His innovative ideas of user-friendly software for the Macintosh changed the design and functionality of software interfaces created for computers. The Macintosh's interface allowed people to interact easily with computers, because they used a mouse to click on objects displayed on the screen to perform some function. The Macintosh got rid of the computer command lines that intimidated people from using computers. After resigning from Apple Inc., Jobs would continue challenging himself to develop computers and software for education and research by starting a new company that would eventually develop the NextStep computer.

Early History

Steven Paul, was an orphan adopted by Paul and Clara Jobs of Mountain View, California in February 1955. Jobs was not happy at school in Mountain View, so the family moved to Los Altos, California, where Steven attended Homestead High School. His electronics teacher at Homestead High School, Hohn McCollum, recalled he was "something of a loner" and "always had a different way of looking at things."

After school, Jobs attended lectures at the Hewlett-Packard electronics firm in Palo Alto, California. There he was hired as a summer employee. Another employee at Hewlett-Packard was Stephen Wozniak a recent drop out from the University of California at Berkeley. An engineering whiz with a passion for inventing electronic gadgets, Wozniak at that time was perfecting his "blue box," an illegal pocket-size telephone attachment that would allow the user to make free long-distance calls. Jobs helped Wozniak sell a number of the devices to customers.

In 1972, Jobs graduated from high school and registered at Reed College in Portland, Oregon. After dropping out of Reed after one semester, he hung around campus for a year, taking classes in philosophy and immersing himself in the counterculture.

Early in 1974, Jobs took a job as a video game designer at Atari, Inc., a pioneer in electronic arcade recreation. After several months working, he saved enough money to adventure on a trip to India where he travelled in search of spiritual enlightenment in the company of Dan Kottke, a friend from Reed College. In autumn of 1974, Jobs returned to California and began attending meetings of Wozniak's "Homebrew Computer Club." Wozniak, like most of the club's members, was content with the joy of electronics creation. Jobs was not interested in creating electronics and was nowhere near as good an engineer as Woz. He had his eye on marketability of electronic products and persuaded Wozniak to work with him toward building a personal computer.

Wozniak and Jobs designed the Apple I computer in Jobs' bedroom and they built the prototype in the Jobs' garage. Jobs showed the machine to a local electronics equipment retailer, who ordered twenty-five. Jobs received marketing advice from a friend, who was a retired CEO from Intel, and he helped them with marketing strategies for selling their new product. Jobs and Wozniak had great inspiration in starting a computer company that would produce and sell computers. To start this company they sold their most valuable possessions. Jobs sold his Volkswagen micro-bus and Wozniak sold his Hewlett-Packard scientific calculator, which raised $1,300 to start their new company. With that capital base and credit begged from local electronics suppliers, they set up their first production line. Jobs encouraged Wozniak quit his job at Hewlett-Packard to become the vice president in charge of research and development of the new enterprise. And he quit his job to become vice president. Jobs came up with the name of their new company Apple in memory of a happy summer he had spent as an orchard worker in Oregon.

Apple Computer

Jobs and Wozniak put together their first computer, called the Apple I. They marketed it in 1976 at a price of $666. The Apple I was the first single-board computer with built-in video interface, and on-board ROM, which told the machine how to load other programs from an external source. Jobs was marketing the Apple I at hobbyists like members of the Homebrew Computer Club who could now perform their own operations on their personal computers. Jobs and Wozniak managed to earn $774,000 from the sales of the Apple I. The following year, Jobs and Wozniak developed the general purpose Apple II. The design of the Apple II did not depart from

Apple I's simplistic and compactness design. The Apple II was the Volkswagon of computers. The Apple II had built-in circuitry allowing it to interface directly to a colour video monitor. Jobs encouraged independent programmers to invent applications for Apple II. The result was a library of some 16,000 software programs.

For the Apple II computer to compete against IBM, Jobs needed better marketing skills. To increase his marketing edge he brought Regis McKenna and Nolan Bushnell into the company. McKenna was the foremost public relations man in the Silicon Valley. Nolan Bushnell was Jobs's former supervisor at Atari. Bushnell put Jobs in touch with Don Valentine, a venture capitalist, who told Markkula, the former marketing manager at Intel, that Apple was worth looking into. Buying into Apple with an investment variously estimated between $91,000 and $250,000, Markkula became chairman of the company in May 1977. The following month Michael Scott, who was director of manufacturing at Semi-Conductor Inc., became president of Apple. Through Markkula, Apple accumulated a line of credit with the Bank of America and $600,000 in venture capital from the Rockefellers and Arthur Roch.

Quickly setting the standard in personal computers, the Apple II had earnings of $139,000,000 within three years, a growth of 700 percent. Impressed with that growth, and a trend indicating an additional worth of 35 to 40 percent, the cautious underwriting firm of Hambrecht and Quist in cooperation with Wall Street's prestigious Morgan Stanley, Inc., took Apple public in 1980. The underwriters price of $22 per share went up to $29 the first day of trading, bringing the market value of Apple to $1.2 billion. In 1982 Apple had sales of $583,000,000 up 74 percent from 1981. Its net earnings were $1.06 a share, up 55 percent, and as of December 1982, the company's stock was selling for approximately $30 a share.

Over the past seven years of Apple's creation, Jobs had created a strong productive company with a growth curve like a straight line North with no serious competitors. From 1978 to 1983, its compound growth rate was over 150% a year. Then

IBM muscled into the personal computer business. Two years after introducing its PC, IBM passed Apple in dollar sales of the machines. IBM's dominance had made its operating system an industry standard which was not compatible with Apple's products. Jobs knew in order to compete with IBM, he would have to make the Apple compatible with IBM computers and needed to introduce new computers that could be marketed in the business world which IBM controlled. To help him market these new computers Jobs recruited John Sculley from Pepsi Cola for a position as president at Apple. Jobs enticed Scully to Apple with a challenge: "If you stay at Pepsi, five years from now all you'll have accomplished in selling a lot more sugar water to kids. If you come to Apple you can change the world."

Jobs in 1981 introduced the Apple III, which had never fully recovered from its traumatic introduction, because Apple had to recall the first 14,000 units to remedy design flaws, and then had trouble selling the re-engineered version. Another Apple failure was the mouse-controlled Lisa, announced to stockholders in 1983. It should have been a world beater, because Lisa was the first personal computer controlled by a mouse which made it have a user-friendly interface, but had an unfriendly price of $10,000. The worst thing about Apple's development of computers was they lacked coherence. Each of Apple's three computers used a separate operating system.

Jobs designed the Macintosh to compete with the PC and, in turn, make Apple's new products a success. In an effort to revitalize the company and prevent it from falling victim to corporate bureaucracy, Jobs launched a campaign to bring back the values and entrepreneurial spirit that characterized Apple in its garage shop days. In developing the Macintosh, he tried to re-create an atmosphere in which the computer industry's highly individualistic, talented, and often eccentric software and hardware designers could flourish. The Macintosh had 128kb of memory, twice that of the PC, and the memory could be expandable up to 192kb. The Mac's 32-bit microprocessor did more things and out performed the PC's 16-bit microprocessor. The larger concern of management concerning the Macintosh was not IBM compatible. This caused an uphill fight for Apple in trying to sell Macintosh to big corporations that where IBM territory. "We have thought about this very hard and it ought to be easy for us to come out with an IBM look-alike product, and put the Apple logo on it, and sell a lot of Apples. Our earning per share would go up and our stock holders would be happy, but we think that would be the wrong thing to do," says Jobs. The strengths of Macintosh design was not memory, power, or manipulative ability, but friendliness, flexibility, and adaptability to perform creative work. The Macintosh held the moments possibility that computer technology would evolve beyond the mindless crunching of numbers for legions of corporate bean-counters. As the print campaign claimed, the Macintosh was the computer "for the rest of us."

The strategy Jobs used to introduce the Macintosh in 1984 was radical. The Macintosh, with all its apparent vulnerability, was a revolutionary act infused with altruism, a technological bomb-throwing. When the machine was introduced to the public on Super Bowl Sunday it was, as Apple Chairman Steve Jobs described it, "Kind of like watching the gladiator going into the arena and saying–Here it is." The commercial had a young woman athlete being chased by faceless storm-troopers who raced past hundreds of vacant eyed workers and hurled a sledgehammer into the image of a menacing voice. A transcendent blast. Then a calm, cultivated speaker assured the astonished multitudes that 1984 would not be like 1984. Macintosh had entered the arena. That week, countless newspapers and magazines ran stories with titles like "What were you doing when the '1984' commercial ran?"

Jobs' invocation of the gladiator image is not incidental here. Throughout the development of the Macintosh, he had fanned the fervour of the design team by characterizing them as brilliant, committed machines. He repeatedly clothed both public and private statements about the machine in revolutionary, sometimes violent imagery, first encouraging his compatriots to see themselves as outlaws, and then target the audience to imagine themselves as revolutionaries. Jobs, like all those who worked on the project, saw the Macintosh as something that would change the world. Jobs described his Macintosh developing team as souls who were "well grounded in the philosophical traditions of the last 100 years and the sociological traditions of the 60's. The Macintosh team pursued their project through grueling hours and against formidable odds. A reporter who interviewed the team wrote: "The machine's development was, in turn, traumatic, joyful, grueling, lunatic, rewarding and ultimately the major event in the lives of almost everyone involved".

The image Jobs wanted the public to have of the Macintosh was young, wears blue jeans, and lives in an 80's version of the 60's counterculture. Macintosh was impatient, uncomfortable, and contemptuous of everything that was conventional or hierarchical. He/she was both creative and committed, believing strongly that his/her work ultimately matters. Even if we counted beans for a living, we secretly saw ourselves as Romantic poets. Jobs approach in developing the Macintosh was like the history of telephones. When the telegraph became popular for communication a century ago, some people suggested putting a telegraph machine on everyone's desk, but everyone would have had to learn Morse code. Just a few years later Alexander Graham Bell filed his first patents for the telephone, and that easy-to-use technology became the standard means of communication. "We're at same juncture; people just are not going to be willing to spend the time learning Morse code, or reading a 400-page manual on word processing. The current generation of personal computers just will not last any longer. We want to make a product like the first telephone. We want to make mass market appliances. What we are trying to develop is a computer that can do all those things that you might expect, but we also offer a much higher performance which takes the form of a very easy-to-use product."

As the Macintosh took off in sales and became a big hit, John Sculley felt Jobs was hurting the company, and persuaded the board to strip him of power. John Sculley tried to change the discipline of the company by controlling costs, reducing overhead, rationalizing product lines to an organization that some in the industry called Camp Runamok. Sculley came to the conclusion that "we could run a lot better with Steve out of operations," he says. [Gelman and Rogers, 1985, p. 46] Jobs tended to value technological "elegance" over customer needs which is a costly luxury at a time of slowing sales. And Jobs' intense involvement with the Macintosh project had a demoralizing effect on Apple's other divisions.

Jobs was exiled to an office in an auxiliary building that he nicknamed "Siberia." Jobs says he did not get any assignments and gradually found that important company documents no longer landed on his desk. He told every member of the executive staff that he wanted to be helpful in any way he could, and he made sure each had his home phone number. Few ever called back. "It was very clear there was nothing for me to do," he says, "I need a purpose to make me go."

He soon came to believe that he would find no purpose within Apple. In July, Sculley had told security analysts in a meeting that Jobs would have no role in the operations of the company "now or in the future." When Jobs heard of the message he said, "You've probably had somebody punch you in the stomach and it knocks the wind out you and you cannot breathe. The harder you try to breathe, the more you cannot breathe. And you know that the only thing you can do is just relax so you can start breathing again."

The Next Step

Jobs sold over $20 million of his Apple stock, spent days bicycling along the beach, feeling sad and lost, toured Paris, and journeyed on to Italy. It was not until late August that he began to catch his breath. Then Jobs thought back on his experience at Apple. Though he is not an engineer, he felt his greatest talent had been spearheading development of new products. Jobs also recalled with special pride that he had helped introduce personal computers into education. To collect his thoughts one day, he took up pen and paper and began to write down the things that were important to him. Along with the development of the

Macintosh, he listed three educational projects he had launched: Kids Can't Wait, Apple Education Foundation, and the Apple University Consortium.

Inspiration came at the beginning of September 1985 when he had lunch with Paul Berg, a Nobel laureate in biochemistry at Stanford University. Paul Berg explained to Jobs the time consuming trial and error experiments carried out to extract DNA. Jobs asked whether Berg had ever thought of speeding up these experiments by simulating them on a computer. Berg said most universities did not have the necessary computers and software. "That's when I started to really think about this stuff and get my wheels turning again," says Jobs.

On September 12, 1985 Steve rose in the board meeting and said in a flay, unemotional voice, "I've been thinking a lot and it's time for me to get on with my life. It's obvious that I've got to do something. I'm thirty years old." Offering to resign as chairman, Steve said he intended to leave the company to start a new venture to address the higher education market. The company Jobs envisioned would have sales reaching $50 million annually in a few years and would not be competitive with Apple, only complementary, and that he would take with him only a handful of personnel. John Sculley said, "all of us want you to reconsider your decision to resign from the board. Apple would be interested in buying 10 percent of your new company." Jobs told the board he would think about it and tell them his decision the upcoming Thursday.

That Thursday Jobs went into Sculley's office and handed him a piece of paper with all five employees that would leave with him. The employees were Rich Page, an Apple Fellow and one of the company's most important engineering designers, Daniel Lewin, the marketing manager for higher education business, Bud Tribble, the manager of software engineering for Macintosh, Susan Barnes, senior controller for U.S. sales and marketing, and George Crow, an engineering manager with vast Macintosh experience. Together, they knew Apple's internal schedules, costs, focus of next products, schedule of when Apple would introduce them, how they would be used, and which individuals and universities Apple would work with to ensure their success. The board authorized Sculley to begin litigation on the basis that Steve allegedly made plans for the new company while serving as Apple's chairman, and that Steve falsely represented his company and intentions to the board.

A Software Company

After leaving Apple, Jobs' new revolutionary ideas were not in hardware but in software of the computer industry. In 1989 Jobs tried to do it all over again with a new company called NextStep. He planned to build the next generation of personal computers that would put Apple to shame. It did not happen. After eight long years of struggle and after running through some $250 million, NextStep closed down its hardware division in 1993. Jobs realized that he was not going to revolutionize the hardware. He turned his attention to the software side of the computer industry.

In 1994, Jobs felt there is a lot of money in developing an object-oriented industry that would fix the problems companies have in developing software. The corporate developers are going to fuel the object revolution because they know they have a giant problem that needs to be solved in software development, and PC makers are doing less to serve the needs of software developers. Jobs said, "Our primary mission is to establish NextStep as a leading operating system in the Nineties." Now, Jobs' NextStep will revolutionize the computer industry by its operating system software which incorporates a hot technology. It's called object-oriented programming (OOP), and OOP lets programmers write software in a fraction of the usual time. Jobs felt OOP is the solution to corporation's problems of wasting money to develop software because OOP serves as a blue print to develop programs like blue prints for constructing a building. Jobs thought the OOP paradigm will have a great effect on the production of software like the effect the industrial revolution had on manufactured goods. "In my 20 years in the industry, I have never seen a revolution as profound as this. You can build software literally five to ten times faster, and that software is more reliable, easier to maintain, and more powerful," says Jobs.

Jobs felt software programs have gone bigger, more complicated, and much more expensive to produce. Object-oriented programming changes that by allowing gigantic, complex programs to be assembled like Tinker toys. Programmers will use pre-assembled chunks of code to build 80 percent of their program thus saving an enormous amount of time and money.

The criticism Jobs received from building the NextStep computer was that he failed in trying to build a second computer empire. Jobs' goal was to produce a NextStep computer for $3,000 that would land on the desk of every college student. In designing the NextStep computer, he ignored the demands of the computer market. Even his own experts were saying: "Keep in touch with the intended customers and avoid the pitfall of anaerobic isolation; do not assume that the customers will pay any price to secure the latest computer technology; ease the way for customers to adopt a new standard by providing software and hardware bridges that help connect older machines to the new ones." According to developers, he disregarded every one of these lessons when he launched NextStep computer.

In mid 1989, after long delays which Jobs was never blamed for, NextStep finally introduced a $7,000 monochrome system. The system had no floppy disk, virtually no useful software applications, and a slow magneto-optical disk. When the NextStep computer was introduced, the academic world and corporate America rejected it. In the end, only about 50,000 NextStep machines were ever built, and in February 1993 Jobs announced that NextStep would stop producing hardware and focus all its energy on the NextStep operating system. The operating system was promised to run on a wide variety of platforms.

Jobs recruited an Englishman, Peter Van Cuylenburg, age forty-four as his number two person in NextStep to help promote the NextStep computer and organize the company's management. The company's management had decimated. In the past few months virtually all of NextStep's vice presidents had quit. Van Cuylenburg said the quitting of vice presidents was due to his own toughness. He said, "I've put pressure on the company, and not everyone was willing or able to accept it. NextStep had too many vice presidents when I arrived, so Jobs and I decided to eliminate some."

Jobs and Cuylenburg planned on releasing NextStep software to run on other companies' computers by the fall of 1993. NextStep did release a version of NextStep's operating system for PC's equipped with Intel's 486 microprocessor. Still, the market did not fully accept NextStep's operating system over OS/2 or Microsoft DOS.

NextStep had also talked with Hewlett-Packard, Sun, and others about licensing NextStep to run on their machines. But these companies thought it was a ridiculous idea, because NextStep is trying to acrimoniously compete against them in hardware. Cuylenburg admits that the scenario makes sense only if NextStep's hardware business is small enough that the major players do not see NextStep's computers as a threat.

Jobs felt NextStep is moving slowly but surely to being a software company that makes great reference hardware. That is NextStep will have a machine that provides a benchmark of quality. The NextStep operating system will be in a three-way race for the object-oriented operating system of the Nineties against Microsoft's Cairo project and Apple's and IBM's joint venture.

Considering that object-oriented software has become the key to NextStep's future, it is ironic that Jobs committed the company to it almost by accident. When NextStep introduced its first machine, the Cube, in 1988, it was incompatible with existing computers. These computers had virtually no software to run on them. Jobs urgently needed outside software developers to write programs for the Cube. He found the basis for his operating system in Carnegie Mellon University software called Mach, which happened to use object-oriented programming. Jobs' goal was not to ease programmers' lives;

he just wanted to get some programs written and shrink-wrapped pronto so he could sell his NextStep computers.

NextStep squeezed its way into the field of being a good platform for companies to build object oriented programming through a review done by CKS Partners. CKS Partners is a San Francisco advertising agency founded by a bunch of old Apple colleagues. Jobs NextStep advertising agency needed help in promoting the NextStep, because it boasted about the computers hardware disk storage and processing chip technology, but gave no compelling reason for businesses to buy a NextStation. Jobs called on his old friends at CKS Partners to help his advertising agency out. CKS conducted focus groups of Fortune 500 managers in charge of information systems. They came up with the report there was little perception in the marketplace about NextStep. But important information came from a number of hard-core information systems geeks. They had discovered NextStep made it much easier and faster for companies' in-house programmers to customize software to handle important parts of their businesses. Rather than start from scratch, programmers using object oriented programming can do much of the job by looking in a library of pre-existing software modules.

This was a good report to have about the qualities and benefits of using a NextStep computer, because if companies were to read analysts' report on NextStep computers in a computer magazine. The companies could reduce the time in developing software packages by having a pre-existing library full of code already written to handle specific operations. And NextStep provides an easy platform to create libraries, maintain, and integrate the code in a object oriented programing environment. The companies would see a solution to the problem of spending too much time and money in building software applications. Software developers could reduce their time in finding errors and maintaining its software, because object oriented design allows a nice encapsulated structure, information hiding, and communication between modules through messages.

A company O'Connor & Associates, a Chicago options and futures firm, claims its engineers can write a complex trading program in three months with NextStep versus over two years on a Sun workstation. Corporate managers who ventured into using NextStep computers told NextStep, "You guys have one of the best products ever, but you do not even know it and you're not trying to sell it to us." Jobs recalls himself, "Companies came to us and said, "You're idiots, you just do not get it." Now that NextStep knew companies in the real world could solve problems faster with NextStep computers. NextStep needed to advertise better how their computers' performance and benefits could make companies more productive. So Next went to compare their system against their number one competitor Sun computer. The Company commissioned a study by management consultants Booz Allen and Hamilton that showed that corporate programmers worked two to nine times faster on NextStep machines, than on Suns and others. When Sun World magazine gave its highest rating not to a Sun machine but to the NextStation Turbo machine, a NextStep advertisement proclaimed: *NEXTSTEP CAST SHADOW OVER SUN.*

From the review reports the company's sales have gone up, but NextStep has been forced to turn to its Japanese partner for cash infusions. Canon originally invested $100 million in 1989 and added another $10 million to $20 million in 1991 before extending that $55 million credit line last July in 1992. Canon held an 18% equity stake. Industry analysts said that the Japanese are increasingly scrutinizing their investment. The heat is on for NextStep to start producing high marginal returns from selling their NextStep products.

Jobs thought Next can survive as a software company when he attacks his old enemy Apple and IBM. He did not think the Apple-IBM linkup will work: "Apple has a thousand software engineers, who have realized that Taligent is their enemy." If Apple adopts IBM's Taligent software, Jobs explained that they would be out of job. Instead, he argued, if Apple will stick with its System 8, under development in-house, leaving IBM as Taligent's primary advocate in the marketplace. This would leave

IBM in a bad position. Jobs admits that Microsoft has "market power" and sees Cairo as his main competitor.

Jobs felt his NextStep machines are going to be in high demand, because once businesses figure out how to use object oriented programming to solve most of their design problems. Businesses are going to buy NextStep computers to run the object oriented platform, because the businesses have money and will pay big money for things that will save them money or give them new capabilities.

Jobs thought the advantages of NextStep software compared to its rival Microsoft is its ability to design programs in an object oriented design. Jobs perceived Microsoft Windows as a bad development environment. And Microsoft does not have any interest in making it better, because the fact that it's really hard to develop applications in Windows plays to Microsoft's advantage. Microsoft developed their software; so, companies cannot have small teams of programmers writing word processors and spreadsheets, because it might upset their competitive advantage. Jobs states that NextStep will become the preferred platform for businesses to develop software. Therefore NextStep software will out-compete Microsoft in programming languages used to develop applications.

Jobs thought object oriented programming will allow small companies to build libraries containing already built coded modules. These libraries will allow programmers to incorporate pre-existing modules to perform specific operations in their code. This type of programming technique will reduce time a programmer has to spend on writing code. Therefore less time spent on a project the company saves money. Since the library code has already been tested, programmers using the pre-existing code in their programs have fewer errors. Less errors to fix in a program means less time spent on the program which saves the company money. Jobs said NextStep software will literally let three people in a small business out perform what 200 people at Microsoft can do. Corporate America has a need to find a solution to their problems. Jobs felt NextStep software can save companies a lot of money or make them a lot of money.

If companies do not presently invest into object oriented programming, their developing technique will cost them large amounts of money. Later, when they try to re-organize their software system they fuel the object revolution.

Jobs believed Microsoft has not transformed itself into an agent for improving things or a company that will lead the next revolution in software development. Jobs has become very concerned because he sees Microsoft competing very fiercely to put a lot of companies out of business. This is hurting innovation in the computer industry. Jobs felt the computer industry needs an alternative to Microsoft's software in computer systems. He hoped people will turn to NextStep software.

How did the creation of Apple and NextStep develop Steve Jobs' managing skills? Jobs has been criticized as America's roughest, toughest, most intimidating bosses. Ever since Steve Jobs founded Apple Computer when he was 21, the meditating computer mogul was known as the terrible infant of Silicon Valley. Now, as head of NextStep, the 38-year-old Jobs is no longer an infant, but according to those who have worked with him, he still is terrible.

Many colleagues described Jobs as a brilliant man who can be a great motivator and positively charming. At the same time his drive for perfection is so strong that employees who do not meet his demands are faced with blistering verbal attacks that can eventually burn out even the most motivated of people. Jobs pushed his workers to the heights of unethical work conditions. In the late 1980's, two NextStep engineers had been slaving nights and weekends for 15 months to meet an important and impossible deadline for a new state-of-the-art chip. No one had ever designed such a thing before, and the strain was incredible. At a weekend off-site meeting Jobs publicly and viciously berated them before the entire company for not working faster, even after all their effort they put into building the chip. Out of pride they finished the project, but one of them quit soon thereafter. A NextStep employee describes his attitude: "You've been on it a week, and you're supposed to be brilliant.

So what have you done? That's why so many people are afraid of him."

Jobs' drive for perfection often lead him to be ignorant to other people's ideas. One ex-employee recalls how Jobs was demanding that, on principle, he would often reject anyone's work the first time it was shown to him. To cope with this unreasonableness, workers deliberately presented their worst work first, saving their best for a subsequent presentation, when it could have a better chance of satisfying the boss's expectations. Several employees felt Jobs is going through a major personality change and becoming much more of a consensus manager and team player.

Steve Jobs, a college dropout who experimented with drugs and Eastern religions before turning to computer design was an unlikely candidate to have become the prototype of America's computer industry entrepreneur. The accomplishments Steve Jobs had on the computer industry while at Apple was introducing the personal computer. Jobs was a bonafide visionary, who created the personal computer, Apple, in his garage. The Apple changed people's view on operations a computer could perform. From computers performing bean counter operations and federal taxes to executing individual's personal business operations. Jobs led a hardware revolution by reducing the size of computers to small boxes.

His development of the Macintosh re-introduced Xerox's innovative idea of user-friendly interface using a mouse. The Macintosh used a windows interface which contained picture-like icons representing a function or a program to be executed. The user would use a mouse to move a cursor onto the icon and press a mouse button to execute the function or program. Companies witness the success of the Macintosh's user-friendly interface and copied its style to develop their software. Jobs, in the nineties, will try to lead another revolution in software development for corporate developers to use the OOP paradigm to solve the massive time and money problems it takes to develop software.

GEORGE DE MESTRAL

One lovely summer day in 1948, a Swiss amateur-mountaineer and inventor decided to take his dog for a nature hike. The man and his faithful companion both returned home covered with burrs, the plant seed-sacs that cling to animal fur in order to travel to fertile new planting grounds. The man neglected his matted dog, and with a burning curiosity ran to his microscope and inspected one of the many burrs stuck to his pants. He saw all the small hooks that enabled the seed-bearing burr to cling so viciously to the tiny loops in the fabric of his pants. George de Mestral raised his head from the microscope and smiled thinking, "I will design a unique, two-sided fastener, one side with stiff hooks like the burrs and the other side with soft loops like the fabric of my pants. I will call my invention 'velcro' a combination of the word velour and crochet. It will rival the zipper in its ability to fasten."

Mestral's idea met with resistance and even laughter, but the inventor 'stuck' by his invention. Together with a weaver from a textile plant in France, Mestral perfected his hook and loop fastener. By trial and error, he realized that nylon when sewn under infrared light, formed tough hooks for the burr side of the fastener. This finished the design, patented in 1955. The inventor formed Velcro Industries to manufacture his invention. Mestral was selling over sixty million yards of Velcro per year. Today it is a multi-million dollar industry.

Not bad for an invention based on Mother Nature.

Velcro and Trademarks

Today you cannot buy velcro because VELCRO is the registered trademark for the Velcro Industries' product. You can purchase all of the VELCRO brand hook and loop fasteners you need. This illustrates a problem inventors often face. Names can become generic terms. Many words used frequently in everyday language were once trademarks, for example: escalator, thermos, cellophane and nylon. All were once trademarked names and only the trademark owners could use the name with a product. When names become generic terms, the U.S. Courts can deny exclusive rights to the trademark.